Synthese Library

Studies in Epistemology, Logic, Methodology, and Philosophy of Science

Volume 468

The aim of *Synthese Library* is to provide a forum for the best current work in the methodology and philosophy of science and in epistemology, all broadly understood. A wide variety of different approaches have traditionally been represented in the Library, and every effort is made to maintain this variety, not for its own sake, but because we believe that there are many fruitful and illuminating approaches to the philosophy of science and related disciplines.

Special attention is paid to methodological studies which illustrate the interplay of empirical and philosophical viewpoints and to contributions to the formal (logical, set-theoretical, mathematical, information-theoretical, decision-theoretical, etc.) methodology of empirical sciences. Likewise, the applications of logical methods to epistemology as well as philosophically and methodologically relevant studies in logic are strongly encouraged. The emphasis on logic will be tempered by interest in the psychological, historical, and sociological aspects of science. In addition to monographs *Synthese Library* publishes thematically unified anthologies and edited volumes with a well-defined topical focus inside the aim and scope of the book series. The contributions in the volumes are expected to be focused and structurally organized in accordance with the central theme(s), and should be tied together by an extensive editorial introduction or set of introductions if the volume is divided into parts. An extensive bibliography and index are mandatory.

Vincent Grandjean

The Asymmetric Nature of Time

Accounting for the Open Future and the Fixed Past

Springer

Vincent Grandjean
Faculty of Philosophy
University of Oxford
Oxford, UK

Published with the support of the Swiss National Science Foundation. Grant no: 10BP12_21 3555/1

ISSN 0166-6991 ISSN 2542-8292 (electronic)
Synthese Library
ISBN 978-3-031-09762-1 ISBN 978-3-031-09763-8 (eBook)
https://doi.org/10.1007/978-3-031-09763-8

This Springer imprint is published by the registered company Springer Nature Switzerland AG
The registered company address is: Gewerbestrasse 11, 6330 Cham, Switzerland

Acknowledgments

I would like to express my sincere gratitude to those Professors, colleagues, and friends who helped me to shape my ideas over the past few years with insightful comments, relevant suggestions or hard questions. I am thinking in particular of Jean-Pascal Anfray, Joshua Babic, Jiri Benovsky, Juan Pablo Bermúdez-Rey, Corine Besson, Robin Bianchi, Philipp Blum, Romain Bourdoncle, Gaétan Bovey, Guillaume Bucchioni, Muriel Cahen, Claudio Calosi, Alessandro Cecconi, Simon-Pierre Chevarie-Cossette, Lorenzo Cocco, Fabrice Correia, Damiano Costa, Daniel Deasy, Alexandre Declos, William Doehler, Jérôme Dokic, Esa Díaz-León, Filipe Drapeau Vieira Contim, Paul Egré, Kit Fine, Benoit Gaultier, Bastien Gauchot, Richard Glauser, Lauréline Grandjean, Ghislain Guigon-Sell, Jean-Baptiste Guillon, Salim Hireche, Kathrin Koslicki, Uriah Kriegel, Raphaël Künstler, Baptise Le Bihan, Frédéric Lelong, Dan López de Sa, Marco Marabello, Stéphane Marbeau, Anna Marmodoro, Olivier Massin, Tim Maudlin, Giovanni Merlo, Anne Meylan, Robert Michels, Cyrille Michon, Kristie Miller, Kevin Mulligan, Paolo Natali, Benjamin Neeser, Frédéric Nef, Judith Notter, Maria Nørgaard, Donnchadh O'Conaill, Matteo Pascucci, Oliver Pooley, Grégoire Quevreux, Antoine Rebourg, Sebastien Richard, Alexander Roberts, Sven Rosenkranz, Pierre Saint-Germier, Maria Scarpati, Alexander Skiles, Daniel Schulthess, Antoine Taillard, Claudine Tiercelin, Lisa Vogt, Antoine Vuille, and Christian Wüthrich.

I also have special thoughts for Otávio Bueno, editor of the *Synthese Library*, for his encouragement, and his continuous support throughout this project.

Work on this book was funded by the Swiss National Science Foundation (SNSF), under the Postdoc.Mobility Fellowship *Time Regained: Reconciling Phenomenological and Physical Perspectives on Time* (P500PH_206602 / 1).

Contents

Chapter 1
Introduction

Abstract In this introduction, my aim is threefold. First, I situate the present book in the vast landscape of the philosophy of time. Against powerful traditions of thought, I argue that a perspicuous philosophical account of time can only be obtained through the reconciliation of the manifest image of the world and contemporary science. Second, I introduce the book's primary objective, namely to account for the intuitive asymmetry between the 'open future' and the 'fixed past', and contrast it with some current forms of skepticism brought by science. Third, I outline the three-phase structure of the book – characterization, modeling, reconciliation – and expose the main expected results.

1.1 Metaphysics of Time: The Science of Reconciliation

When we think of time, no accurate definition comes to mind. As Augustine famously puts it: "[w]hat then is time? If no one asks me, I know; if I want to explain it to a questioner, I do not know" (2006: 242). However, some features of time some are experienced quite vividly. First, time flows uniformly and universally. It indeed seems that time passes at the same speed everywhere, i.e. no matter where we are located. For example, if two friends separate – one starts living in the Swiss Alps, the other in Paris – and they meet up again years after, we think that time will not have elapsed more for one than for the other; one will not have lived longer than the other. Second, it seems that 'our present' extends throughout the whole universe. In particular, it makes perfect sense for us to call a friend who is visiting New York to ask him what he is doing *now*. Regardless of the jet lag issue, we assume that our friend and ourselves can have *simultaneous* activities. For example, while I am writing this introduction, he is visiting the Guggenheim Museum. Third, and perhaps most importantly, it seems that there is a difference between the future and the past: the former is *open* while the latter is *fixed* (or *closed*). For example, whereas we think that there are things we can do to affect how the future will unfold (e.g., making a significant donation to an NGO, acting in an environmentally responsible manner), we think that the past is beyond our control ('what is done is done').

V. Grandjean, *The Asymmetric Nature of Time*, Synthese Library 468,
https://doi.org/10.1007/978-3-031-09763-8_1

Although these features of time play roles of varying levels of importance in our epistemic life, they are all pieces of the 'manifest image of the world', where by 'manifest' one should understand 'the world as it appears to us' (cf. Sellars, 1962: 37–38). This is supported by various empirical works (cf., for instance, Latham et al., 2020a, b, 2021a, b).

The problem is that science, especially contemporary physics, contests these features of time, and hence contradicts our intuitions. First, Albert Einstein understood – a century before we had clocks precise enough to measure it – that time passes more slowly in some places, more rapidly in others.[1] To use the previous example, if two friends separate – one starts living in the Swiss Alps, the other in Paris – and they meet up again years after, the one who has stayed in Paris will have had a shorter life than the one who has stayed in the Alps. This is a consequence of General Relativity (GR), which posits that a large mass – such as the Earth – warps spacetime and thus slows down time in its vicinity. This so-called 'time-dilation' effect is more important in the plains than in the mountains, because the plains are closer to the center of the gravitational field.[2] Second, as Special Relativity (SR) has stated, there is no unique present, for there is no absolute relation of objective simultaneity: two spatially separated events may be simultaneous for one observer, and temporally distant for another observer. It therefore makes no sense to call a friend who is visiting New York to ask him what he is doing *now* (or while I am writing this introduction), for there is no privileged moment that constitutes *the* present. Third, since the 'block universe' view of time (which seems implied by SR) is *isotropic* (spacetime has no intrinsic direction) and the fundamental laws of physics are *time-reversal invariant* (they do not distinguish the future-direction from the past-direction), it appears that any intuitive asymmetry in time is at best a non-fundamental phenomenon (that plausibly results from how matter is contingently distributed through spacetime), and at worst an illusion (due to the peculiar way our brains interact with the world). This leads us to think that our intuition that the future is open while the past is fixed is not physically grounded, at least at the fundamental level.

Given this, it seems that any philosopher who aims at providing a perspicuous account of time has to deal with at least two conflicting concepts of time: on the one hand, *phenomenal time*, of which we possess direct knowledge through mental awareness and, on the other hand, *physical time*, of which we have indirect knowledge. Whereas phenomenal time is embedded in the perspective of a particular encounter with the world, physical time is *inferred* from important results obtained in contemporary physics (especially SR and GR). Once the conflict between these two

[1] Of course, this terminology is metaphorical: in Relativity, velocity is not absolute. A more accurate way to introduce this idea would be to say that the geometry of spacetime is *influenced* by the distribution of matter.

[2] Besides gravitational potential differences between locations, Special Relativity (SR) indicates that time dilation may be due to two clocks having a velocity relative to each other. For example, if, instead of sending two friends to the mountains and the plains respectively, we ask one of them to stay still while the other walks around, time will pass more slowly for the one who keeps moving (cf. Rovelli, 2018: 31).

concepts of time has been properly identified, the issue that arises for philosophers is 'How to manage this conflict?'. Roughly, in recent history of philosophy, two rival options have been privileged, though they both issue from a 'conflict-avoidance' strategy: subjectivism and scientism. Subjectivism consists in being exclusively concerned with intuitions to the detriment of science; scientism in abandoning intuitions for the benefit of science. Although these two options have taken on a variety of forms, for our purpose it will suffice to briefly describe each of their two most famous manifestations: *phenomenology* and *the philosophy of inner life* (or *Bergsonism*) on the one hand, and *empiricism* and *naturalized metaphysics* on the other.

First of all, many twentieth-century philosophers have been interested in the phenomenology of temporal awareness. Maurice Merleau-Ponty, for example, was highly critical of scientific conceptions of time. He famously put forward a view of time that agrees more with "[...] the descriptions given by literary men and artists than that given by physicists" (Mays, 1972 : 355). As Merleau-Ponty puts it: "[t]here is more truth in mythical personifications of time than in the notion of time considered, in the scientific manner, as a variable of nature itself, or, in the Kantian manner, as a form ideally separable from its matter" (1962: 422). Although "[...] his remarks apply specifically to the human cultural situation rather than to the scientific context [...]" (Mays, *id*) ("mythical personifications of time" would be regarded as a gross anthropomorphism by physicists), they still inform us about the role that philosophy is, in his view, meant to play. According to Merleau-Ponty, philosophy must concentrate on describing time as it is reflected in human experience, i.e. as the ultimate subjectivity, which "[...] reveals the subject and the object as two abstract moments of a unique structure, namely, *presence*" (1962: 474).

Of course, Merleau-Ponty's view may, in some respects, appear to be caricatural. But the fact remains that phenomenology has, from its earliest beginnings, excluded physical time from philosophy's primary concerns. Edmund Husserl, for example, with his watch-word 'back to the things' argues that philosophy must begin with the phenomena themselves: philosophers must concentrate on describing the authentically given – the contribution of empirical sciences is relegated to the background, even within transcendental phenomenology, which reproaches science for not questioning the foundations of the objectivation of the world. As a result, physical time, which is not adequate to the task of explaining how consciousness experiences temporal objects, is neglected in philosophical inquiry, understood as an epistemological analysis of temporal lived experience. For example, whereas Newtonian time (conceived as an empty container of discrete, atomistic *nows*) can explain the separation of moments in time (e.g., separated tones), it cannot explain the continuity of these moments (e.g., a melody). Such a quantitative view of time must therefore be supplanted by a phenomenological attempt to articulate *how* flowing objects are experienced. As Husserl (1964: 26) makes clear, a perspicuous account of time must explain *how* temporal objects, though composed of distinguishable moments (which can be measured by clocks), are apprehended as a unity (rather than as a convoluted patchwork).

Second, some philosophers have been interested in the notion of *duration* as being a qualitative multiplicity, i.e. a temporal heterogeneity in which "[...] several conscious states are organized into a whole, permeate one another, [and] gradually gain a richer content [...]" (Bergson, 2010: 122). This approach is heir to Henri

Bergson's *philosophy of inner life*, which famously criticizes the scientific spatialization of time: scientists conceptualize time as an ordered arrangement of defined events, rather than as an unceasing flux impressed upon consciousness.[3] Specifically, in the Bergsonian view, space is fundamentally unlike time: space is *homogenous* (in space, things exist separate from alongside each other), *divisible*, and *infinite*, whereas time is *heterogenous* (in time, things interpenetrate and are never completely independent), *indivisible* and *finite*. Space is a homogenous medium, whereas time is a process that emerges continuously in the absolute new.[4] For example, while one often draws a parallel between the ticks of a clock and time itself, Bergson argues that there is 'no point of contact' (other than conventional)[5] between these things: the clock's hands move through space, not time. According to Bergson, duration (or 'real time') is ineffable and can only be grasped through a simple intuition of the imagination; attempts to intellectualize or measure duration are doomed to failure, at least if one seeks to understand reality in its deepest nature. The Bergsonian intuition is, however, not merely a pre-theoretic experience of the world (as described above), but a method that consists in entering into the things (rather than going around them from the outside) to get absolute knowledge. As a result, the common representation of time as a one-dimensional line (on which the present is a point) is, according to Bergson, a pure abstraction: time is mobile and incomplete, and the present consists in the consciousness of our body experiences. Moreover, a scientific concept such as 'spacetime', according to which time is just as much a dimension as any of the spatial ones, is quite alien to Bergson: "[b]asically space and time play opposite roles in his philosophy, which might even be regarded as a dualism of space and time" (Lacey, 1999: 17). Crucially, Bergson's approach should not be confused with phenomenology: whereas phenomenology takes the multiplicity of phenomena to be always related to a *unified* consciousness, Bergsonism takes the immediate data of consciousness to be a *multiplicity*.

An immediate objection that could be raised against this first 'subjectivist' option of managing the conflict between intuitions and science is to observe that the world existed in time before human beings and human consciousness ever appeared on the Earth.[6] For example, it appears that the emergence of the first living organisms

[3] Bergson's criticism of the spatialization of time culminated in his debates with Einstein, and in the book that tentatively grappled with relativistic physics, *Duration and Simultaneity* (1922).

[4] The differences between space and time are detailed in Bergson (1972: 515). It is worth noting that the idea of *infinite* space is abandoned in *Matter and Memory* (1896).

[5] As Bergson puts it: "[w]e substitute [...] for the qualitative impression received by our consciousness, the quantitative interpretation given by our understanding" (Bergson, 2010: 51). Interestingly, Bergson describes the spatialization of time as a social convention, which results from our pragmatistic look at the world – we look at the world from the point of view of action, while philosophy's role is precisely to discard this point of view.

[6] This objection has been taken up by Merleau-Ponty. One reply to it might be, he says, that "[...] every equation in physics, presupposes *our* pre-scientific experience of the world" (1962: 422). In that sense, no physical concept is independent of man and the cultural environment in which he finds himself. However, this reply seems to fail, since one can perfectly conceive things, e.g., possible worlds, without being able to have sense-awareness of them.

temporally follows ocean formation, which implies that time was already passing, say 4 billion years ago. It therefore seems simply wrong to claim that time has no reality beyond our subjectivity. Moreover, philosophers (especially metaphysicians of time) may have more ambition than merely describing *phenomena.* They may want, for instance, to contribute (together with physicists) to the exploration of the fundamental structure of reality and, therefore, to the exploration of physical time. For example, philosophers may aim to account for the notion of 'the passage of time', not conceived as a subjective phenomenon, but rather as a central aspect of the scientific picture of the world. Tim Maudlin, for instance, argues that "[t]he passage of time is an intrinsic asymmetry in the temporal structure of the world, an asymmetry that has no spatial counterpart" (2007: 108-109). Of course, such a *revisionary* (as opposed to *descriptive*)[7] way of undertaking metaphysics must be scientifically well informed. After all, assuming that philosophers and physicists can both investigate physical time, they cannot deliver contradictory truths about it. For these reasons, it seems that dismissing scientific resources from philosophical inquiry is methodologically doubtful.

The second option of managing the conflict between phenomenal and physical time is to ignore our intuitions and to concentrate solely on science. This scientistic option has taken on a variety of forms, but we will focus on two of them: *empiricism* and *naturalized metaphysics*.[8] First, many philosophers of physics tend to think that it is solely empirical science that provides access to the fundamental structure of reality. For instance, Otto Neurath writes that "[t]here is no such thing as philosophy as a basic or universal science alongside or above the various fields of the one empirical science; there is no way to genuine knowledge other than the way of experience; there is no realm of ideas that stands over or beyond experience" (1973: 136). This thought is historically based on some 'verificationist principle', according to which every synthetic statement must be empirically verifiable.[9] As a result, most metaphysical statements are meaningless. For example, empiricists typically believe that a true picture of time could only be arrived at by a study of such things as atomic clocks and physical processes. In this perspective, the way the world is pre-theoretically grasped has no relevance. Against the rationalist tradition, empiricists do not see intuitions as cognitive tools designed to produce guidance toward the truth, but rather as a misleading product of our immediate environment, which

[7] The expression 'descriptive metaphysics' was first coined by Peter Strawson in his book *Individuals* to capture his task to "[...] describe the actual structure of our thought about the world" (2003: 9).

[8] 'Naturalized metaphysics' should be contrasted with 'moderately naturalized metaphysics', as it is defended by Matteo Morganti and Tuomas Tahko (2017), the latter of which seems to partly legitimate the use of intuitions.

[9] This formulation of the verificationist principle is obviously too stringent. For example, it renders all universal generalizations (which are empirically unverifiable) meaningless and, therefore, undermines vast domains of science and reason. That is why some logical positivists (Rudolf Carnap, Otto Neurath, Hans Hahn and Philipp Frank) tried to make the verificationist principle more inclusive (this movement is known as the 'liberalization of empiricism') (cf. Sarkar & Pfeifer, 2006: 83).

must in turn be studied by the precise methods of empirical science.[10] According to empiricism, the role of philosophy is merely to react to scientific discoveries by reshaping traditional concepts, such as time and reality. But, while (at least some) philosophers make use of intuitions, their work is of no value when it comes to discerning the fundamental structure of reality. Of course, this is not to say that there are no difficulties in the scientific theories, but rather that training in philosophy or discussing these difficulties with subjectivist philosophers would not help to solve them. Their work is fundamentally misguided, after all.[11]

Another, though less radical, way of dismissing intuitions is promoted by some metaphysicians of science, who assume that the main role of philosophy is to unify hypotheses and theories that are taken seriously by contemporary science. As James Ladyman and Don Ross famously claim: "[n]aturalized metaphysics must go beyond mere consistency with current science; it must be directly motivated by and in the service of science" (2013: 109). Here the starting point is clearly science; metaphysics comes second and builds upon scientific results. For example, when statistical mechanics was extended to the quantum realm, it was found that in order to obtain empirically adequate statistics, particles of the same kind (e.g., fermions) must be indistinguishable, though they can be counted. This induces a metaphysical response, namely that the Leibnizian Principle of Identity of Indiscernibles is false. Here again, although philosophy contributes to the inquiry into the nature of reality (e.g., in terms of theoretical unification), the use of intuitions is strictly limited. In particular, naturalized metaphysics explicitly excludes any reasoning based on intuitions and *a priori* considerations, for "[…] there is no reason to imagine that our habitual intuitions […] are well designed for science" (Ladyman & Ross, 2007: 2). Such a metaphysical stance should be contrasted with 'Lewisian-style' metaphysics. In short, whereas Lewis takes for granted that abandoning intuitions (e.g., the intuition that causes occur prior to their effects) should be regarded as a cost associated with accepting a metaphysical thesis, naturalists, such as Ladyman and Ross, are not concerned with preserving intuitions at all; they are merely concerned with serving science.

An first objection that could be raised against this scientistic option of managing the conflict is that it rests on a caricature of science. It is wrong to claim that scientists exclusively rely on empirical data and rational methods, at least in the context of discovery. As Jonathan Tallant (2013, 2014) has highlighted, the term 'intuition' is frequently applied by scientists to explanations, pictures, and results. This can be illustrated by an ethnographic study, carried out by Ference Marton and Peter Fensham (1993), which is based on data collected from 83 Nobel Prize-winners, drawn from physics, chemistry, and medicine. This study reveals that most of the

[10] Some studies (e.g., Morris et al., 1995) tend indeed to show that intuitions depend on our ontogenetic makeup and partly on culturally specific learning.

[11] This 'empiricist' view is part of the heritage of logical positivistic work against metaphysics, which aims to show that "[…] all meaningful discourse can be reduced to, or at least rigorously justified by reference to, reports of observations regimented for communication and inference by formal linguistic conventions […]" (Quine, 1966: 151), see also Carnap, 1931.

Laureates (72) say they use scientific intuitions to guide the shape of their inquiries. For example, as Michael Brown[12] declares, "[…] we almost felt at times that there was almost a hand guiding us. Because we would go from one step to the next, and somehow we would know which was the right way to go. And I really can't tell how we knew that, how we knew that it was necessary to move ahead" (cf. Tallant 2014: 295). Of course, one might reply that the use of intuitions by scientists is not the same as that of metaphysicians. For example, one might argue that scientists, unlike metaphysicians, never use intuitions to *justify* their claims. But, even if one grants this difference, it is does not follow that metaphysicians always use mere intuitions to this end, nor that they cannot but do so. In particular, although the term 'intuition' is notoriously vague, one should agree that *both* scientists and metaphysicians may use intuitions at least "[…] in the sense of background assumptions influencing how they interpret (empirical) data […]" (Morganti & Takho, 2017: 2570).

This more realistic picture of science seems to match with *how* Arthur Eddington and John Wheeler introduce physical knowledge. Contrary to Albert Einstein (1936), who claims that "[w]e can view only as miraculous that our sense-experience can be unified by our freely created concepts",[13] Eddington and Wheeler do not believe in miracles. In particular, they do not believe that scientific concepts are freely created by pure thought alone divorced from our actual intuitions. They both emphasize the role of inferences from 'data', 'pointer readings' or 'information' to phenomena and knowledge of physical reality. As Eddington points out, all physical knowledge is 'hypothetico-observational knowledge', which means that science is an inference from good observations (cf. 1939: 49).[14] This view can even be pushed further if we consider, as Alfred North Whitehead does, that some important scientific concepts – such as velocity, acceleration, momentum, and kinetic energy – are unintelligible when disconnected from temporal experience. For example, it seems impossible to define 'velocity' without some reference to the past and the future, which makes essential the importation of this distinction into the physical picture of time. As Whitehead puts it, "[t]his conclusion is destructive of the fundamental assumption that the ultimate facts for science are to be found at durationless instants of time" (1919: 2).

A second objection against the scientistic option is that if our intuitions are indeed a misleading product of our immediate environment, then it should be explained *why* it is so. There is a need to explain *why* people across the countries and the centuries are systematically deceived by their intuitions when apprehending time. Such a 'massive error hypothesis' cannot be put forward without providing any proportionate explanation. For example, given the 'block universe' view of time (which seems implied by SR), the spacetime geometry must have been fixed and settled all at once in some single act of creation. Accordingly, you and I did not

[12] Nobel Prize-winner for Medicine in 1985.

[13] The quote is from Palter (1960: 4).

[14] For more details on the 'hypothetico-observational' conception of physical knowledge, see Weinert, 2017.

come into existence later than dinosaurs or the Big Bang. Rather, if the universe were created at all, then "[...] you and I and all its parts, and all the associated events in which these parts are involved, were created together" (Brogaard, 2000: 345). But if everything is in existence (in some sense of 'is' that is both tenseless and timeless), then the advocate of such a model is faced with the problem of accounting for the common fact that "[...] new perceptions enter and exit our minds in successive fashion at what appear to be always later and later times" (Brogaard *id.*). How does one explain that our cognitive abilities are limited in such a way that we have no access to the future, if there is no difference between the present, the past and the future? And why is it that the times at which we perceive the world appear to be ordered successively? It is not clear that the advocate of the 'block universe' view can account for these facts. Of course, one could reply that potential solutions to these issues are to be found outside the philosophical field (e.g., in psychology). But these kinds of buck-passing answers are clearly unsatisfying: one cannot skip the question as to *why* everybody is mistaken in apprehending time, under the pretense that this could hypothetically be answered by other scientists. Worse still, as Mauro Dorato argues: "[...] the ontology posited by a physical theory should in principle be capable of establishing connections with the world of our experience, for the latter world is the source of *the empirical tests of the theory*" (2008: 54).

A last objection is that, although it should be acknowledged that many things that have seemed obviously true have turned out, upon inspection, to be false (e.g., that the sun rotates around the Earth), it seems that appealing to rational intuitions is *epistemically justified*. Two arguments can be invoked in favor of this claim.

First, what we have learnt from early modern attempts to find absolute certainties is that, unless we are willing to become extreme skeptics, we must allow that "[...] it is reasonable to believe things that seem obviously true, in the absence of special reasons to doubt them; and we must allow this even if the beliefs are admittedly not certainties, and cannot be 'proven' in any interesting sense of the word" (Zimmerman, 2008: 222). This leads to a methodological principle, sometimes referred to as 'the principle of credulity' (cf. Broad, 1939; Swinburne, 1979), that seems reasonable to accept: intuitions must be preserved as long as they are not proven wrong. In that sense, intuitions are 'innocent until proven guilty' and their relinquishing should only be envisaged as a last resort, when our best science leaves no hope of preserving the manifest image of the world.

Second, intuitions have demonstrated success in the past. In particular, there are many examples in which a philosophical theory was found flawed because it contradicted an intuition (especially in specific thought experiments). For example, in epistemology, the 'classical' theory of knowledge was abandoned because, in Gettier cases (1963), we share the intuition that one can have the justified true belief that *p* without knowing that *p*. Moreover, in the philosophy of language, the descriptivist theory of proper names was rejected because, in Kripke cases (1980), we share the modal intuition that, although someone other than the U.S. President in 1970 might have been the U.S. President (e.g., Humphrey might have), no one other than Nixon might have been Nixon. Finally, in applied ethics, the naive utilitarian theory was rejected because, in Thomson cases (1976), we share the intuition that it is *not* morally permissible to take five organs of a healthy person in order to prevent five people from imminent death, though this action would maximize well-being. These three examples – but there are many more – show that intuitions decisively contribute to philosophical progresses in a great variety of fields and, without further arguments, it is hard to see why it would not be the case in the metaphysics of time. Of course, one might emphasize that these intuitions crucially differ from the intuitions we have about the nature of time, as they are not in tension with important scientific results. But, this does not answer the question as to *why*, whereas intuitions seem reliable in such various fields as epistemology, philosophy of language, and applied ethics, they would not be so in the metaphysics of time.

A further criticism could be that the three former cases (Gettier, Kripke, Thomson) are not relevant, as they do not involve any reasoning about the metaphysical structure of the world from our intuitions about the world. For example, the knowledge case differs from metaphysical theorizing in that it crucially rests on some concept that we commonly use: knowledge. Assuming that we are competent with that concept, we can elucidate its content by asking about the conditions under which we would employ it. In brief, we ask people if they would employ it in Gettier case, and we find they would not. So, that provides a reason to suppose that our concept of knowledge (as we commonly use it) is not just justified true belief. Yet, this does not seem like a very robust sense of an appeal to intuition in settling how things are with the world: we merely investigate how we are disposed to use some concept (viz. knowledge), to determine what we mean by it. It therefore seems that there is no analogy with metaphysical theorizing. In reply, two things can be said. First, the three former cases were not intended to provide an analogy, but merely to illustrate the fact that intuitions are widely and successfully used in various domains of philosophy. Second, one can easily think of cases where intuitions have played a decisive role in metaphysical theorizing. Consider, for example, Kit Fine's influential rejection of modal conceptions of essence, according to which "[...] an object [has] a property essentially just in case it is necessary that the object has the property" (1994: 3). The rejection is based on five properties (e.g., 'being a member of singleton Socrates', 'being distinct form the Eiffel Tower') that necessarily belong to Socrates, but that are *intuitively* not essential to Socrates. The five properties thus function as five counterexamples to the classical, modal conception of essence, which is thus explicitly rejected by Fine on the sole basis of intuitions. The above arguments against scientism provide as many reasons to think that our intuitions must not only be explained but also vindicated.

Now, while none of the two 'conflict-avoidance' strategies (subjectivism vs scientism) is fully satisfying, one may wonder whether all hope of solving the conflict between intuitions and science is gone. In that respect, the situation is not altogether desperate, as a third way – perhaps more demanding, but also more interesting – of managing this conflict can be imagined: the way of reconciliation. As Craig Callender puts it: "[w]e seem to have, to echo another debate, an 'explanatory gap' between time as we find it in experience and as we find it in science. Reconciling these two images of the world is the principal goal of philosophy of time" (2008:

339). The main idea behind 'reconciliation' is to bridge the gap between the time of human experience and that of science, by critically analyzing and conceptually improving the way non-physicists think about the nature of time. This requires working in close contact with both pre-theoretic and scientific data and, thereby, to develop a framework in which these data can be articulated non-paradoxically. In order to do so, it is worth following three methodological steps. First, one has to provide a rigorous philosophical characterization of our intuitions, by surveying, for example, the different senses in which 'time passes', 'our present extends throughout the whole universe' and 'the future is open while the past is fixed'. Second, one has to determine which model of the temporal structure of the world is most appropriate to accommodate these intuitions. Third, one has to describe this model in such a way that it meets the main imperatives of our most salient scientific theories. The present book fits into this general philosophical scheme.

Although this scheme places a lot of weight on the vindication of intuitions, this should not overshadow the fact that other factors (e.g., simplicity, parsimony) also play an important role when evaluating a model. In that sense, our main objective is not to arrive, *whatever the cost*, at the model that best fits a set of specific intuitions – such a model might turn out to be complex, unparsimonious, or in tension with our best science, after all. And, clearly, even if intuitions play a crucial role in theorizing, they certainly do not trump all these other factors. Rather, our main objective is to find an equilibration between how well a model coheres with best science, the picture it provides of the world, how parsimonious, simple, and so on the model is, and how well it accords with our intuitions. This methodological precision is intended to avoid the risk of settling on models that are pretty baroque.

From a more general perspective, it may be worth contrasting the methodology promoted in this book with other philosophical attempts to reconcile the manifest image with the scientific image. In that respect, the Canberra Plan, initiated by David Lewis and Frank Jackson (cf. Jackson, 1994, 1998; Lewis, 1970, 1972), seems particularly relevant. This program of philosophical methodology and analysis (which brings together many people who were associated with the Australian National University in Canberra during the 1990s) primarily aims at reconciling a certain account of conceptual analysis with philosophical naturalism. Specifically, the approach can be broken in two steps (cf. Braddon-Mitchell & Nola, 2008: 7). First, we collect together the 'platitudes' concerning the *X* to be analyzed (e.g., colors) – these platitudes may simply be the large number of what we (or experts) can agree are the truths about *X* (e.g., snow is white and normal perceivers in normal conditions have experiences of white caused in them). These agreed-upon platitudes about *X* are expressed in two kinds of terms: the 'outsider' O-terms, i.e. the terms that get their meaning from *outside* the platitudes (e.g., 'perceivers', 'conditions'), and the 'insider' T-terms, i.e. the terms that play a theoretical role specified in the platitudes (e.g., the terms for the colors of objects and experiences). Second, we discover what in our best theory of the world, if anything, plays the theoretical roles spelled out by the T-terms; or, to put it like Braddon-Mitchell and Nola, "[…] what our current best theories tell us there is in the world to serve as realizers of the theoretical roles specified in the platitudes" (2008: 7). Sometimes, nothing in the current sciences can serve as a realizer of the specified role. When that happens, most proponents of the Canberra Plan, who are usually physicalists, conclude that those realizers fail to exist (as it has been the case, for example, with the phlogiston).

Another current approach is more focused on our assertions and practices. First, we examine the conditions under which we use certain terms, and the conditions under which we have certain practices as part of the manifest image. Second, we find what is in the scientific image that explains those assertions and practices. For instance, if we go about talking about colors, and engaging in colors practices, we look at the scientific image and see what explains this: perhaps the fact that there is electromagnetic radiation of a certain range of wavelengths visible to the human eye. We then conclude that this is what it is for there to be colors. As before, there may be cases in which what we learn about what explains our assertions and practices leads us to say that, in fact, we are mistaken: there are no such things. So, here again, perhaps if what explains our use of 'free will' assertions and practices turns out to be very different from what we expected, we come to conclude that there is no free will. Alternatively, perhaps we revise our notion of what free will is, to come in line with what it is that explains our assertions and practices.

These two popular approaches share an important feature: they seek to 'locate' the manifest image *inside* the scientific image, allowing that what we find might both *explain* our having those intuitions as well as *vindicate* the relevant intuitions, or might *explain* our having those intuitions without vindicating them. In general, any program that aims at reconciling the two images should meet the following minimum requirement: it has to explain, by appeal to the scientific image, why things seem as they do according to the manifest image. In that respect, the methodology promoted in this book deserves the label of 'reconciliation program', although it diverges from the two previous approaches in some crucial aspects. First, my methodology is more speculative and exploratory than those described above. As it will become clear in §4, I do not only examine well-established scientific theories to explain and vindicate relevant intuitions, but I also explore nascent theories (e.g., the causal set theory) that, although they do not (yet) strictly belong to our best science, provide a valuable insight into what our best science might look like in the near future. Second, as a consequence of the previous point, the conclusion that my methodology allows me to reach is necessarily more modest than one might expect: all that can be shown is that science (including nascent theories) does not *a priori* rule out the possibility that certain relevant intuitions adequately reflect the structure of the world. Although the modest nature of the conclusion may be disappointing, it should be contrasted with the fact that any conclusion based on well-established scientific theories, at least in fundamental physics, is at best temporary. As it will be detailed in Sect. 4.6, General Relativity and Quantum Mechanics are incomplete; they fail to capture phenomena that combine high energy densities and strong gravitational fields (e.g., the very early universe, and the dynamics of black holes). These theories will therefore 1 day plausibly be replaced by a theory of Quantum Gravity (such as the causal set theory) with superior predictive and explanatory power.

1.2 A Temporal Asymmetry: The Open Future and the Fixed Past

The present book is mainly related to the metaphysics of time and the philosophy of science. It offers a detailed study and a systematic defense of a key intuition we typically have, as human beings, with respect to the nature of time: the intuition that the future is open, whereas the past is fixed. As will be shown in the second chapter, there are many ways in which this intuition may manifest itself. But, as a first approach, it will suffice to think of some future and past events. For example, whereas it seems unsettled whether there will be a fourth world war, it is settled that there was a first world war. In that sense, whereas nothing (in the present or the past) *a priori* predetermines that a large-scale armed conflict will blow up in the future (it is ours to prevent such a disaster!), there is nothing we can do to prevent WWI from having taken place in the past. Likewise, whereas it seems that behaving in an environmentally responsible manner may prevent some animal species from extinction, there is nothing we can currently do to bring back the dodo birds. It is however worth noting that, although the question of human abilities (what we can or cannot do) may inform us on the asymmetry in openness between the future and the past, it would be a mistake to reduce this asymmetry to purely anthropocentric considerations. In this book, one will also consider senses in which the future and the past may respectively be said 'open' and 'fixed' in a world without humans (or before humanity emerged).[15]

Generally speaking, it seems very hard to discard the intuition that the future is open while the past is fixed. For, this intuition, which is to be understood as a pre-theoretic representation (or concept) of the temporal structure of the world, is reflected everywhere in our relationship to the world. We have certain *practices* that are time asymmetric: we plan for the future not the past; we deliberate about the future not the past; we take some past apparently fixed facts as inputs into our deliberation; we act as though we can causally influence the future but not the past; etc. We also have certain *emotions* and *attitudes* that are time asymmetric: we feel regrets about the past, not the future; we anticipate the future, not the past; we generally place greater value on future events over past ones, etc. Finally, we have *observations*, which, around here, are of asymmetric: we have records of the past, but not the future; we observe that entropy increases towards the future and away from the past, etc. All these things are taken as manifestations of a basic intuition that deeply structures our relationship to reality, and that must not only be explained but also vindicated for the reasons set out in Sect. 1.1.[16]

[15] For similar reasons, the question as to whether the future is open should not be confused with the question as to whether humans are free (though these two questions are often associated). In Sect. 2.4, I will present compatibilist arguments, according to which even if the future turns out to be fixed (because determinism is true), there is a sense in which some human actions are free.

[16] This book will not provide an account of how our practices (and similar notions) emerge. Interestingly, however, such an account could both explain, and vindicate, certain of our practices, without thereby being one that would vindicate the future being open according to our pre-theoretic representation of the temporal structure of the world.

However, the main models of the temporal structure of the world do not reflect any asymmetry between the future and the past. According to presentism and eternalism, the future and the past are ontologically on a par. Eternalists hold that both the future and the past exist, while presentists hold that neither the future nor the past exists. In other words, the two main competing models of the temporal structure of the world do not ontologically distinguish the future from the past (either both of them exist or none of them exists). Therefore, neither eternalism nor presentism seems able to account for our basic intuition regarding the nature of time. Of course, one might claim that the asymmetry between the open future and the fixed past does not need to be grounded in the temporal structure of the world, but merely in some natural processes (e.g., the increase of entropy). Nonetheless, one will provide reasons to think that these natural processes are not as important as some philosophers would have us believe for temporal asymmetries. For example, the increase of entropy – which arises from the collective behavior of many microscopic entities – at best postpones the problem: if there is no directedness in fundamental physics, where does the thermodynamic asymmetry in time come from?

This project might be criticized for taking too seriously the intuitive asymmetry between the open future and the fixed past. After all, some arguments taken from science, especially from contemporary physics, have been put forward to show that the asymmetry is at best a non-fundamental phenomenon, at worst an illusion. For example, the 'block universe' view of time, which is *inferred* from the results of SR, does not reflect any asymmetry. It regards reality as a block-like four-dimensional ensemble, lacking a moving present, wherein all times and events are equally real. In other words, the spatiotemporal model favored by contemporary physics does not reflect any difference between space and time that somehow accounts for the fact that whereas there is no here-there space-asymmetry, there should be a past-future time asymmetry. Likewise, the fundamental laws of physics, which are *time-reversal invariant*, do not underpin any asymmetry regarding the nature of time: for every physically allowable sequence of events, the inverse sequence of *time-reversed* events is also physically allowable. For example, if one watches a movie that shows a ball rolling, the fundamental laws of physics cannot tell whether the movie is being projected correctly or in reverse. Therefore, accounting for the asymmetry between the open future and the fixed past as a *fundamental* phenomenon seems to require developing an alternative model of the temporal structure of the world to that favored by contemporary physics, namely a spatiotemporal model that has the *intrinsic resources* to ground the asymmetry. This model will turn out to be a specific version of C. D. Broad's growing block theory of time (GBT).[17]

The final step of the book will be the reconciliation of this alternative model (GBT) with contemporary physics. Although physics cannot settle the debate about

[17] Admittedly, GBT will only provide a metaphysical ground to our intuition, not a complete account of it. This latter task would require further psychological investigations on the emergence of intuitions, which will not be undertaken within this book.

the nature of time,[18] it crucially *informs* and *frames* the debate. As Yuval Dolev puts it: "[physics] has shaped the manner in which we set out to study our world, and our understanding of almost every aspect of it" (2006: 188–189). It therefore seems that no metaphysical contribution to the question of the nature of time can be provided without observing the main imperatives of our best physical theories. In that respect, considerable efforts have to be devoted to the understanding of the main postulates and consequences of both the Special and General theory of relativity, but also to the nascent theories of quantum gravity (which aim to unify GR with the principles of quantum mechanics). The main purpose of this final step is to show that GBT is *expressible* in a relativistic spacetime setting and can, thereby, offer a naturalistic basis to the intuitive asymmetry between the open future and the fixed past. In this perspective, the causal set approach to quantum gravity (CST) will be a matter of great interest, since it allows for the 'coming-into-existence' of events through a discrete stochastic process and might, therefore, underwrite a specific version of GBT. At the end of the day, the time of human experience might turn out to be a more faithful reflection of the time of science than what most philosophers believe.

1.3 Three Main *Desiderata*

The present book aims to satisfy three main *desiderata*: (i) to provide a coherent, non-metaphorical, and metaphysically illuminating characterization of the intuitive asymmetry between the 'open future' and the 'fixed past'; (ii) to determine which model of the temporal structure of the world is most appropriate to accommodate the asymmetry; (iii) to reconcile this model (and hence the asymmetry) with our best physics. Each of these *desiderata*, which echo the three methodological steps introduced in Sect. 1.1, will feature as a separate chapter within the book. The second chapter deals with characterization, the third chapter with temporal models, and the fourth chapter with our best physics. From a broader perspective, satisfying *desiderata* (i), (ii), and (iii) should allow one to obtain a framework within which pre-theoretic and scientific data can be articulated non-paradoxically. This seems required for bridging the gap between the time of human experience and that of science. Let's have a closer look at these three *desiderata*.

First, although the intuition that the future is open and the past fixed is widely shared, it is not a straightforward matter to determine the nature of the asymmetry it reflects. So, in the second chapter, I review the various philosophical ways of characterizing the asymmetry in order to account for our intuition. In particular, I wonder whether the asymmetry should be characterized in a *perspectival* way (it merely reflects how humans interact with the world) or in a *substantial* way (it reflects how

[18] One reason is that one can describe many structures that everywhere look *locally* like possible spacetime structures, while no physicist would consider them to be representations of possible ways spacetime can be; see for instance the non-orientable Möbius two-dimensional spacetime (cf. Maudlin, 2012: 157).

the world *truly* is). It is worth noting that perspectival and substantial characterizations do not exactly serve the same function: the latter aim both at explaining and vindicating our intuitions, whereas the former are more about vindicating our having certain practices. My conclusion is that substantial characterizations are more promising, as they are the only ones that offer a *fundamental* explanation (i.e. not causal or thermodynamic) of the asymmetry reflected by our intuition. Specifically, assuming physical indeterminism (i.e. the doctrine that the future history of the world is *not* nomologically necessitated by its current history), I argue that the asymmetry between the open future and the fixed past is a kind of worldly unsettledness that should be characterized in ontological terms: there being facts of the matter about what happened, but not about what will happen. This characterization, which stems from the 'no fact of the matter' account of openness, will be shown to be superior to the alternatives in explanatory power, intelligibility, and in how it coheres with interesting senses of openness.

Second, while the above characterization requires the asymmetry to be reflected in the temporal structure of the world, neither of the two main competing models – eternalism and presentism – ontologically distinguishes the future from the past. Eternalists hold that both the future and the past exist, whereas presentists hold that neither the future nor the past exists. Thus, neither eternalism nor presentism can avail itself for the 'no fact of the matter' account of openness, while keeping the past fixed.[19] For instance, although the 'no fact of the matter' account of the open future is available to presentists, they cannot endorse it without acknowledging that the past is open in the very same sense. In the third chapter, I therefore propose to reject both of these models in favor of a 'growing block theory' of time (GBT), which is naturally seen as better designed to accommodate the asymmetry. Against powerful traditions, I do not introduce GBT as a hybrid between eternalism and presentism, but as real alternative: GBT is the only *asymmetric* theory of time (i.e. necessarily sometimes the spatiotemporal structure it describes is not reflection invariant) that accepts *Temporal Becoming* (i.e. the creation of new things in the present). Finally, I address the so-called 'epistemic objection', which purports to show that GBT leads to absolute skepticism about *where* we are temporally located, by appealing to the continued existence of bare particulars and to introspective knowledge. I take this occasion to explain (i) *how* existence in the past should be conceived and (ii) *why* it differs sharply from existence in the present.

Third, GBT is often criticized for conflicting with important results of contemporary physics. For example, Hilary Putnam (1967) and Wim Rietdijk (1966) have both argued that the view according to which the future is unreal requires an objective notion of absolute simultaneity, while such a notion is rejected by the theories of relativity. Yet, since one aims to end up with a scientifically coherent account of time, it has to be shown that GBT is *expressible* in a relativistic spacetime setting.

---

[19] Of course, this does not necessarily prevent presentists from claiming that some statements about the past are true (or false) *now*. But this claim requires a relaxed conception of the grounding requirement on tensed truths, according to which the present truth of a given statement does not require it to be grounded in how things located in the present are (cf. Sect. 2.9).

In the fourth chapter, I reply to the Putnam-Rietdijk argument by putting forward a certain approach to quantum gravity, which is formulated in terms that naturally underwrite GBT. In particular, I argue that the causal set approach to quantum gravity (CST), when enriched with the 'classical sequential growth' dynamics (CSG), depicts reality more as a 'growing being' than as a 'static thing'. This result apparently disproves the widespread claim that fundamental physics undermines any attempt to defend an open-future view. In short, it is possible to be both a scientific realist *and* a defender of the view that the future is genuinely open. Finally, I move from science to science fiction and show that GBT is, in principle, compatible with some scenarios such as time-travel.

As I hope to have indicated above, despite all that has already been accomplished, much work still remains to be done in reconciling the manifest image of the world and contemporary science. In what follows, I intend to contribute to this undertaking by offering my own thoughts on the challenges faced by philosophers who aim to bridge the gap between the time of human experience and that of science.

Chapter 2
How Is the Asymmetry Between the Open Future and the Fixed Past to Be Characterized?

Abstract A basic intuition we have regarding the nature of time is that the future is open whereas the past is fixed. However, although this intuition is largely shared, it is not a straightforward matter to determine the nature of the asymmetry it reflects. So, in this chapter, I survey the various philosophical ways of characterizing the asymmetry in openness between the future and the past in order to account for our intuition. In particular, I wonder whether the asymmetry is to be characterized in semantic, epistemic, anthropocentric, physical, modal, metaphysical or ontological terms. I conclude that an ontological characterization of the asymmetry is to be preferred, since it is superior to the alternatives in explanatory power, intelligibility, and in how it coheres with interesting senses of openness.

2.1 Introduction

A basic intuition we have regarding the nature of time is that there is a difference between the future and the past: the former appears to be open and the latter appears to be fixed (or closed). This intuition manifests itself in various ways. First, whereas we think of the future as *partially unsettled* (e.g., it is settled that I will die someday, but it is unsettled whether the first astronaut to go to Mars will be a woman), we think of the past as *fully settled* (e.g., it is settled that Napoleon lost at Waterloo, that dinosaurs are extinct animals). Second, whereas we think that there are things we can do to affect how the future will unfold (e.g., making a significant donation to an NGO, acting in an environmentally responsible manner), we think that there are not

This chapter is based on Grandjean (2020, 2021a) and Grandjean and Pascucci (2021), but contains new material.

V. Grandjean, *The Asymmetric Nature of Time*, Synthese Library 468,
https://doi.org/10.1007/978-3-031-09763-8_2

things we can do to affect how the past unfolded ('what is done is done').[1] Third, whereas our future experiences are of great concern to us, we attach little importance to our past experiences (e.g., we would prefer, other things being equal, that personal bad events be in the past rather than in the future).[2] Fourth, and perhaps more radically, whereas we may only wonder *how* the past unfolded (e.g., 'What happened to John Kennedy?'), we may wonder *whether* the future will unfold (e.g., 'Will reality continue beyond tonight?').

The intuition of an asymmetry in openness between the future and the past is so deeply ingrained in our manifest image of the world that it seems hopeless to do without. We decide, we create, we remember, we regret. The first two attitudes presuppose an open future, while the two latter ones presuppose a fixed past. For example, when it comes to forming beliefs about what we remember or regret, we explore our mental life, i.e. we consult our memory and records, since these latter attitudes are epistemically constrained by the information we may collect about what happened to us. By contrast, when we want to know what we will decide or create, i.e. when we try to predict the outcomes of such pending processes, we do not gather psychological evidence or records, since any information we might obtain will be overridden by the processes themselves. We rather let these processes run their courses; they are almost guaranteed to produce true beliefs (cf. Ismael, 2016: §6). This suggests that whereas our attitudes towards the past depend on the traces it left on our mental life, the future *partially depends* (either directly, or in an attenuated manner) on our decisions and our creations. As, for instance, Mauro Dorato puts it: "[o]ur actions can give a (cosmically negligible) contribution to bring [the future] about" (2008: 56). It therefore seems that, unlike the past, the future cannot be regarded as more fixed than the processes in which we are currently engaged.[3]

However, although the intuition of an open future and a fixed past is largely shared, it is not a straightforward matter to determine the nature of the asymmetry it reflects. So, in this chapter, I survey the various philosophical ways of characterizing the asymmetry in order to account for our intuition. In particular, I discuss the question whether the asymmetry is to be characterized as *semantic* (the principle of bivalence applies to statements about the past but not to future contingents), *epistemic* (we can know much more about the past than we can know about the future), *anthropocentric* (we can affect what will happen, but not what happened), *physical*

[1] Geach (1973: 211) rightly points out that, contrary to appearances, the locution "what is done is done" is not tautological. When we say "what is done is done", we are not saying that if Napoleon's army lost the battle of Waterloo then it lost the battle of Waterloo, but rather that if it is true at some earlier time that Napoleon's army lost the battle of Waterloo then nobody (not even God, according to Aquinas) will be able at any later time to bring it about that Napoleon's army did not lose it.

[2] This phenomenon, which has been shown by a series of studies in psychology, is known as 'the temporal value asymmetry' (or 'TVA') (cf. Caruso et al., 2008). It will be discussed in Sect. 5.3.

[3] Of course, I do not claim that our intuitions regarding genuine decisions and creations provide an argument in favor of the open future. After all, as Popper points out, "[a] man may well believe that he is acting deliberately, and of his own free choice, when in fact he is acting under the influence of suggestion, or of compulsion, or of drugs" (1988: 1). What I claim is rather that if we take the future to be fixed, then our decisions and creations cannot be regarded as less fixed.

(the world is fully deterministic with regard to the past but not to the future), *modal* (if the present were different, the future would be different, but the past would remain as it in fact is),[4] *metaphysical* (whereas the past is singular, there are many alternative future states such that the world fails to specify which ones obtain) or *ontological* (there being facts of the matter about what happened, but not about what will happen). I conclude that, although many of these characterizations may contribute to a global understanding of the phenomenon, an ontological characterization of the asymmetry is to be preferred, since it is superior to the alternatives in explanatory power, intelligibility, and in how it coheres with interesting senses of openness. Of course, this does not exclude the possibility that other accounts – e.g., those involving epistemic, causal, counterfactual or entropic asymmetries – may (alongside various cognitive mechanisms) shed light on related phenomena such as, for example, our having certain practices regarding the future. But this is not the primary concern of this chapter. What matters here is simply to obtain the best characterization of a crucial aspect of our pre-theoretic representation of the structure of the world, namely our intuition that the future is open and the past fixed.

In this respect, the overall picture is perhaps more fragmented than has been suggested so far. For, the various accounts that will be discussed below do not aim to explain the very same thing. Whereas some of these accounts are explicitly intended to *characterize* what it is for the future to be open, and the past fixed (i.e. they are attempts to capture our representation of openness and fixity), some others seem better conceived as taking our relevant practices seriously and asking: 'What could explain our having these practices?'. In particular, the 'perspectival' accounts that resort to asymmetries of knowledge (or causation) are clearly of the second kind: they aim less at characterizing the openness of the future than at explaining why we believe the future is open when it is not (or only in a 'perspectival' sense). Also, these accounts vindicate our having certain practices (e.g., our deliberating about the future and not the past, and using information about the fixed past in the deliberation), without thereby vindicating that the future is open in some other (more substantial) sense. In this respect, the 'perspectival' accounts must crucially be contrasted with the 'substantial' accounts (e.g., the metaphysical and ontological accounts) that aim at vindicating the future being open in the sense in which we might have a pre-theoretic representation of its openness.

[4] This view *a priori* precludes determinism, i.e. the thesis that at any time the world has exactly one possible future, given the past and the laws of nature.

2.2 The Failure of Bivalence

It has become increasingly popular to claim that the asymmetry in openness between the future and the past is an asymmetry with respect to whether some statements[5] about the future and the past have a classical truth-value, i.e. are either true or false.[6] Arguably, this claim is originally to be found in Aristotle's *De Interpretatione*, chap. 9. In this book, Aristotle seems to capture the asymmetry between the open future and the fixed past by a semantic claim: future contingents (i.e. statements about the future that, even if they should be either true or false *now*, their present truth-value would anyway not be *predetermined* by the present or the past)[7] are neither true nor false, whereas statements about the past are either true or false.[8] In other words, it might be argued that, for Aristotle, the 'open future' view amounts to the failure of Bivalence (which states that all meaningful statements are either true or false) when applied to future contingents. Aristotle writes:

> In the case of that which is or which has taken place, propositions, whether positive or negative, must be true or false. Again, in the case of a pair of contradictories, either when the subject is universal and the propositions are of a universal character, or when it is individual, as has been said, one of the two must be true and the other false; whereas when the subject is universal, but the propositions are not of a universal character, there is no such necessity. […] *When the subject, however, is individual, and that which is predicated of it relates to the future, the case is altered* (Aristotle, 2014: 31 [my emphasis]).

As an example, Aristotle famously considers the case of a sea-battle, which has since served as the focal point for most of the philosophical discussions concerning the open future. In Aristotle's picture, a sea-battle "[…] must either take place to-morrow or not, but it is not necessary that it should take place to-morrow, neither is it necessary that it should not take place, yet it is necessary that it either should or should not take place to-morrow". Now assuming that Bivalence holds unrestrictedly and, therefore, that future contingents, such as ⌜There will be a sea-battle tomorrow⌝, are either true or false at the time they are asserted, it seems, as Aristotle puts it, that "[...] everything takes place of necessity and is fixed". In particular, assuming that the statement ⌜There will be a sea-battle tomorrow⌝ is true *now*, it seems that tomorrow cannot be peaceful, because "[...] that of which someone has

[5] Following Peter Strawson (1950), I define 'statements' as 'uses of sentences'. It is sentences that have meaning, but statements that have truth-values and between which logical relations hold.

[6] Cf. Markosian, 1995, Tooley, 1997, Macfarlane, 2003, Diekemper, 2004, and Curtis & Robson, 2016.

[7] In that sense, even if some future contingents should presently be true (or false), their present truth-value would anyway "not be made inevitable" by facts that are, strictly speaking, facts about what goes on in the present or what went on in the past (cf. Correia & Rosenkranz, 2018: 110).

[8] The reason why I do not simply speak of 'statements about the future that are metaphysically neither necessary nor impossible' is that they do not all trigger intuitions of unsettledness. For instance, ⌜I will die someday⌝ is a future tensed-statement that is metaphysically neither necessary nor impossible, but the proposition it expresses is intuitively settled.

said truly that it will be, cannot fail to take place; and of that which takes place, it was always true to say that it would be" (Aristotle, 2014: 31–32).

Although there are various interpretations of Aristotle's writing on this and other issues, most commentators[9] agree that this Aristotelian argument – commonly referred to as 'the fatalist argument' – can be reconstructed as follows:[10]

1. Either it is true that there will be a sea-battle tomorrow or it is false that there will be a sea-battle tomorrow.
2. If it is true that there will be a sea-battle tomorrow, then it is true *now* that there will be a sea-battle tomorrow, and likewise, if it is false that there will be a sea-battle tomorrow, then it is false *now* that there will be a sea-battle tomorrow.
3. If it is true *now* that there will be a sea-battle tomorrow, or false *now* that there will be a sea-battle tomorrow, then how tomorrow is (at least with respect to sea-battles) is settled by how the present is.
4. Therefore, how tomorrow is (at least with respect to sea-battles) is settled by how the present is.
5. Since we were dealing with an arbitrary event at an arbitrary future time, how the future is in all respects is settled by how the present is.

Given that step (5) is the denial of the claim that the future is open in any respect whatsoever, the fatalist argument leads Aristotle to conclude that the openness of the future[11] cannot be preserved without excluding future contingents from the scope of Bivalence, i.e. without denying (1). That is presumably why some philosophers[12] go a step further by *identifying* the 'open future' view with the claim that Bivalence does not hold for future contingents. In this sense, these philosophers do not consider the failure of Bivalence as merely *implied* by the openness of the future (if the future is open, then future contingents are neither true nor false), but claim that the openness of the future *is nothing but* the non-bivalence of future contingents (the future is open iff future contingents are neither true nor false). Here is, for example, what Markosian writes about the open future:

> Let us agree on some terminology. To say, with regard to some time, t, that the future is open at t is to say that there are some propositions about the future relative to t that are, at t, neither true nor false. To say that the future is closed at t is to deny this, i.e., to say that every proposition about the future relative to t is, at t, either true or else false (1995: 96).[13]

---

[9] Cf. Haack, 1974, Markosian, 1995, Barnes & Cameron, 2009, Besson & Hattiangadi, 2013, Le Poidevin, 2013, and Curtis & Robson, 2016.

[10] This particular reconstruction of the fatalist argument can be found in Barnes & Cameron (2009: 292).

[11] Aristotle (2014: 32) speaks rather of the "potentiality of the future", i.e. the potential for incompatible possibilities to eventuate.

[12] Cf. Markosian, 1995, Macfarlane, 2003, Diekemper, 2004, Curtis & Robson, 2016.

[13] Of course, the fatalist argument – even if it is accepted – does not force us to adopt Markosian's terminology, i.e. to define the openness of the future as the failure of bivalence, *but the converse is not true*. The rejection of the fatalist argument, especially of step (3), undermines Markosian's terminology.

So, according to Markosian, tomorrow is open (at least with respect to sea-battles) iff (i) it is not true *now* that there will be a sea-battle tomorrow and (ii) it is not false *now* that there will be a sea-battle tomorrow. And, of course, since the openness of the future is not confined to potential sea-battles, those who adopt Markosian's terminology will typically claim that *all* future contingents (at least as defined above) are neither true nor false. For example, as I write these lines, it may be claimed that it is neither true nor false that the first astronaut to go to Mars will be a woman, a cure for cancer will be discovered by the year 2115, and Federer's grandson will also become a famous tennis player. Of course, everyone is free to define the openness of the future as they want, especially in such a way that it is analytic that the future is only open if future contingents are non-bivalent. But such a definition might appear unsatisfying for at least three reasons that I present now.

First of all, the claim that the future is open is meant to capture some basic intuitions we have regarding the nature of time (partial unsettledness of the future, power over what will happen, etc.), and if we identify the 'open future' view with the claim that Bivalence does not hold for future contingents then we risk simply missing the point. The non-bivalence of future contingents has indeed nothing to do with the way we commonly think of time. That may in particular be revealed by our pragmatic assessments concerning the correctness and the incorrectness of statements about how things would turn out (cf. Besson & Hattiangadi, 2013).

As a first example, consider our current assertions of future contingents; it seems natural to regard some of them as correct. There would, for instance, be nothing *prima facie* problematic in someone's asserting that ⌜I will brush my teeth tonight⌝. However, it is generally taken to be a necessary condition of an assertion's being correct that it is true; so that if an assertion of a proposition is believed to be not true, it will not be assessed as correct (cf. Grice, 1989). Given this, Markosian seems compelled to conclude either that the future is fixed (at least with respect to assertions of future contingents that are assessed as correct), or that we are massively mistaken when we feel pragmatically justified in asserting future contingents. Neither of these options is acceptable.

As a second example, consider our past predictions about how things would turn out; it seems natural to regard some of them as correct *retrospectively*. For instance, while we may think that it is *now* open whether or not there will be a sea-battle tomorrow, once tomorrow comes and there is indeed a sea-battle, we are not only inclined to think that it is true *now* that there is a sea-battle, but we are also inclined to think that this reveals that yesterday's prediction that there would be a sea-battle *was* correct. However, it is hard for anyone who takes openness to consist in (or even to imply) the non-bivalence of future contingents to agree with this. In particular, when Markosian considers predictions made in the past about how things would be at a time that is now the present, he seems forced into saying that they were neither true nor false and – given the orthodox account of assertion – that they could not have been correct. After all, assuming that the openness of the future consists in the failure of Bivalence for future contingents, it must be concluded that while there is *now* a sea-battle, yesterday's prediction that there would be a sea-battle today was not correct, because it was open how things would turn out (cf. Macfarlane, 2003:

324–325). That is unacceptable. As Gilbert Ryle writes: "[i]t is an unquestionable and very dull truth that for anything that happens, if anyone had at any previous time made the guess that it would happen, his guess would have turned out correct" ([1953] 2015: 19).

A possible answer to this objection lies in adopting a *relativist semantics* whereby tensed claims have a truth-value only relative to a context of assessment (cf. Macfarlane, 2003, 2008). The main idea is that the very same claim – the prediction made on Monday that there will be a sea-battle on Tuesday, say – lacks a truth-value when assessed relative to the time of utterance (Monday), but is either true or false when assessed relative to the time whose goings on the claim is making a prediction about (Tuesday). However, although this answer seems to reconcile the openness of the future with our intuition about retrospective assessments, it has an important theoretical cost: the rejection of the *absoluteness of utterance-truth* (i.e. the orthodox assumption according to which the truth-value of an utterance does not depend on some context of assessment) (cf. Evans, 1985: 349–350).

A second reason why we might think that identifying the 'open future' view with the failure of Bivalence for future contingents is unsatisfying has to do with the fatalist argument. Although this argument seems to be valid, some of its steps – especially (3) – can be disputed. It is not clear whether the bivalence of future contingents rules out the openness of the future. Many philosophers[14] defend the view that the openness of the future is compatible with the bivalence of future contingents and, therefore, that the statement ⌜There will be a sea-battle tomorrow⌝ can be either true or false *now* without settling how the future will be. For example, Barnes and Cameron (2009, 2011) reject step (3) of the fatalist argument. They claim that the move from 'if it is either true or false *now* that *p*' to 'it is *now* settled that *p*' relies on a mistaken assumption, namely that if a statement has a truth-value then it is settled that it has *that* truth-value. For Barnes and Cameron, it can be settled that a statement has a truth-value (either truth or falsity), without it being settled which truth-value this statement has. Specifically, it is settled that ⌜There will be a sea-battle tomorrow⌝ is either true or false, but it is neither settled that this statement is true nor that it is false. This claim leads Barnes and Cameron to conclude that the bivalence of future contingents can be reconciled with a peculiar kind of open future (expressed in terms of *metaphysical indeterminacy*) and, therefore, that the fatalist argument must be rejected.

Likewise, Correia and Rosenkranz (2018) reject step (3) of the fatalist argument. They argue that it rests on too strong a conception of the so-called 'grounding requirement on tensed truths' (i.e. the requirement according to which tensed truths do not 'float free', but are grounded in reality). Their idea is that, although the truth-value of a future contingent must be grounded in reality (i.e. in what exists and how things that exist are), it does not need to be grounded in how things located in the present or past (of now) are or have been. In particular, "[…] the present truth of a

---

[14] Von Wright, 1979, Greenough, 2008, Prawitz, 2009, Barnes & Cameron, 2009, 2011, Torre, 2011, Besson & Hattiangadi, 2013, Cameron, 2015, Todd, 2016, Correia & Rosenkranz, 2018.

statement about how, at some future time, things will be, might well be said to be, at that future time, going to be grounded by things being that way" (2018: 108). For instance, supposing that ⌜There will be a sea-battle tomorrow⌝ is true *now*, there will be a sea-battle tomorrow such that it will explain *why*, one day before, the statement ⌜There will be a sea-battle tomorrow⌝ was true. Now, since tomorrow's sea-battle is not *predetermined* (nothing there is or was, in conjunction with how it is or was makes it inevitable), Correia and Rosenkranz conclude that future contingents can *now* have a truth-value without the future being bound to be a certain way; so they reject step (3) of the fatalist argument.

A last example is given by Todd's Russellian approach to future contingents. According to Todd (2016), there is a crucial connection between the debate about 'The present King of France' (cf. Russell, 1905, 1957; Strawson, 1950) and the debate about the open future: just as everyone denies that there exists 'the present King of France', so the open futurist denies that there exists an actual future. Taking this parallel seriously, the Strawsonian view of statements such as ⌜The present King of France is bald⌝ and the Aristotelian view of future contingents look very similar: they both take the relevant statements to be neither true nor false. Therefore, just as there is a Russellian alternative to the Strawsonian view, according to which statements such as ⌜The present King of France is bald⌝ are all simply *false*, so there must be a Russellian alternative to the Aristotelian view, according to which future contingents are all simply *false*. Todd develops this latter alternative, which rests on a very simple principle: a future tensed-statement is true iff what it says happens in the actual future. But given that there is no actual future (as open futurists contend), then nothing happens in the actual future and any future-tensed statement is *false*. Now, since a statement such as ⌜There will be a sea-battle tomorrow, or there will not be a sea-battle tomorrow⌝ is not an instance of $p \vee \neg p$ (even if it can easily seem that it could not fail to be true, at least if the 'open future' view is false),[15] Todd concludes that the 'open future' view can be endorsed without any violation of the classical logical principles of bivalence and Excluded Middle; so he also rejects step (3) of the fatalist argument.[16]

Finally, a third reason why Markosian's attempt to define the openness of the future seems flawed relates to the important costs generated by the denial of Bivalence. In particular, future contingents cannot be presented as counterexamples to Bivalence without specifying what logic and semantics one ought to assume when reasoning about the open future. Two options are generally retained to model truth-value gaps: (i) assuming a three-valued treatment of truth-functional

[15] According to Todd, the real instance of $p \vee \neg p$ is this: ⌜There will be a sea-battle tomorrow, or it is not the case that there will be a sea-battle tomorrow⌝. (This strategy has originally been devised by Russell himself to preserve classical logic).

[16] As will be made clear in the next sections, I do not accept the reasons of Barnes & Cameron or Todd for which they reject step (3) of the fatalist argument. In particular, Todd's approach is incompatible with Ryle's common-sensical view according to which at least some future contingents are correct (cf. p. 20). These examples are only mentioned to illustrate the variety of reasons one may have for rejecting step (3) of the fatalist argument.

connectives, (ii) assuming supervaluationism. However, both of these options have well-known drawbacks.

For example, on either Kleene's or Łukasiewicz's three-valued logics, $F\varphi \vee \neg F\varphi$ is neither true nor false when $F\varphi$ is a future contingent, which is clearly undesirable: even if it may be *now* unsettled whether or not there will be a sea-battle tomorrow, it should anyway be settled that either there will be, or there will not be a sea-battle tomorrow (cf. Aristotle, 2014: 32, Prior, 1953). This problem cannot be solved in terms of three-valued semantics if the logic is truth-functional, i.e. if the truth-value of any proposition always depends entirely on the truth-values of its parts. As has been argued by Arthur Prior (1953: 326), changing the truth-tables to something different from Łukasiewicz's model would be useless. As long as the model is truth-functional, it is clear that the two disjunctions $F\varphi \vee \neg F\varphi$ and $F\varphi \vee F\varphi$ will have the same truth-value (cf. Øhrstrøm & Hasle, 2020: §4.1). This is not satisfactory, since $F\varphi \vee \neg F\varphi$ is clearly true, whereas $F\varphi \vee F\varphi$ is undetermined, given that $F\varphi$ is undetermined.

With supervaluationism no comparable problem arises. The non-bivalent semantics it affords underwrites all theorems of classical logic, including every instance of $F\varphi \vee \neg F\varphi$. According to this theory, a statement is true at a time *t* just in case it is supertrue at *t*, i.e. just in case it is true at *t* on all histories that include *t*. Likewise, a statement is false at *t* just in case it is superfalse at *t*, i.e. just in case it is false at *t* on all histories that include *t*. In all other cases, a statement is neither true nor false. In particular, future contingents – that are true on some future histories, false on others – are neither true nor false, where this must be understood not in the sense that future contingents have a third truth-value (as in Kleene's or Łukasiewicz's logics), but that they *lack* a truth-value. It is thus common to define future contingents as 'gappy' because supervaluationism, contrary to Kleene's or Łukasiewicz's three-valued logics, allows for truth-value gaps. However, since supervaluationism retains the Excluded-middle while it rejects Bivalence, it has to abandon the Tarski biconditional ('φ' is true iff φ)[17] and, therefore, the disquotational property of truth, which could turn out to be unacceptable. Tim Williamson, for instance, writes that "[h]ow much more there is to the concept of truth than the disquotational property is far from clear, but in most contexts truth is assumed to be at least disquotational, whatever else it is or is not" (1994: 162).[18]

Moreover, as Williamson (1994: 151) points out, the following rules of inference are classically valid, yet they may fail in a language with a supervaluational semantics:[19]

[17] Such a proof has been provided by Haack (1974: 67).

[18] A possible solution to this objection would be to say, as Thomason (1970: 273) does, that Tarski biconditional holds only as a consequence ($\varphi \vDash$ true 'φ') and not as an implication (so that for some φ, $\nvDash \varphi \rightarrow$ true 'φ'). However, this solution violates the deduction theorem and, therefore, leads to non-classicality.

[19] The failure of rules [2] and [3] is already noted in Fine (1975b) and Machina (1976), respectively.

[1]	From $\Sigma, \varphi \vDash \psi$ infer $\Sigma \vDash \varphi \rightarrow \psi$	*Conditional proof*
[2]	From $\Sigma, \varphi \vDash \psi$ infer $\Sigma, \neg\psi \vDash \neg\varphi$	*Contraposition*
[3]	From $\Sigma, \varphi \vDash \psi \wedge \neg\psi$ infer $\Sigma \vDash \neg\varphi$	*Indirect proof*
[4]	From $\Sigma, \varphi \vDash \sigma$ and $\Sigma, \psi \vDash \sigma$ infer $\Sigma, \varphi \vee \psi \vDash \sigma)$	*Proof by cases*

It might therefore be argued that, since classical logic and semantics are vastly superior to the alternatives required by the denial of bivalence in "[…] simplicity, power, past success and integration with theories in other domains" (Williamson, 1994: 186), they should as far as possible be preserved, and so Markosian's characterization of the asymmetry – which leads to non-classicality – should be rejected.

In a nutshell, since we aim (i) to capture some basic intuitions about the nature of time, (ii) to question the fatalist argument, and (iii) to possibly retain classical logic and semantics, it seems wrong to begin with the supposition that the 'open future' view amounts to the failure of Bivalence for future contingents. Rather, we should ask how best to understand our basic intuitions regarding the nature of time, while leaving open the possibility that the failure of Bivalence may end up being a non-logical *consequence* of the 'open future' view. In other words, it seems that the costly rejection of Bivalence is not *definitional* of the open future and, if needed, should only be motivated by our best understanding of our intuitions about time. For these reasons, it seems preferable to look for another, presumably non-semantic, way of characterizing the asymmetry, and ultimately assess whether it might be reconciled with an unrestricted application of Bivalence.

2.3 A Reflection of Our Ignorance

It is often assumed that the asymmetry between the open future and the fixed past is merely an epistemic phenomenon: we can know much more about the past than we can know about the future. For example, if we want to know who won the Nobel Prize for literature last year, or whether John Kennedy was killed on a Tuesday, we can consult our memory or look it up in a book. After all, we find ourselves in a world with plenty of information about the past. By contrast, we have no records of who will win the Nobel Prize for literature next year, no books in which we can look up whether the first astronaut to go to Mars will be a woman. Of course, we can make guesses about how the future will be, but our guesses are "spotty" and "provisional" (Ismael, 2016: 140). In this sense, the asymmetry in openness between the future and the past is not among the 'fundamental features' of reality. It only reflects the fact that whereas we are in a position to gain a wide knowledge of the past, the future remains largely unknowable to us.

At first sight, this understanding of the asymmetry as an epistemic and therefore non-fundamental feature of reality accords well with important results of contemporary physics. For example, the theory of relativity seems to imply a 'block universe' view of time,[20] in which the asymmetry does not arise. According to this view, the block universe extends from the Big Bang to the end of time if there is one, or indefinitely, if there is not. It represents all times as equal parts of reality, i.e. without making any fundamental asymmetric distinction between them. Just as spatial places (e.g., Greenwich Village, Plaça de Catalunya) exist, despite not being *here* (in Switzerland), so too past and future times exist, despite not being *now* (in 2022).[21] In other words, the spatiotemporal model favored by contemporary physics does not seem to reflect any difference between space and time that somehow accounts for the fact that whereas there is no here-there space-asymmetry, there should be a past-future time asymmetry.

Likewise, the fundamental laws of physics, which are time-reversal invariant (insofar as the positions of particles are concerned),[22] do not underpin any asymmetry regarding the nature of time. For example, the laws of classical electrodynamics – since they entail that whatever motions particles can execute, they can execute backward – fail to capture any temporal asymmetry: "[...] the unbreaking of glass can be no less in accord with the laws of Maxwellian electrodynamics than the breaking of glass is, and the spontaneous heating of soup can be no less in accord with Maxwellian electrodynamics than its spontaneous cooling is, and the coming of youth can be no less in accord with Maxwellian electrodynamics than its passing is" (Albert, 2000: 15). After all, for a glass to break, or for a soup to be spontaneously heated is just for their constitutive particles to assume certain particular sequences of positions. And, since every sequence of positions $S_I \ldots S_F$ (which is in accord with the laws of classical electrodynamics) admits the inverse sequence of time-reversed positions $S_{F'} \ldots S_{I'}$ (which is also in accord with these laws), it turns out that classical electrodynamics (as well as most post-Newtonian theories, such as the Special and General theory of relativity, and the standard interpretations of quantum mechanics) makes no significant difference between the future and the past (at least with respect to their fundamental laws).[23]

Of course, one might emphasize that there are exceptions. In particular, one might claim that classical thermodynamics is *partially* governed by its Second Law, which states that the total entropy of any isolated system (i.e. no outside influences and no leakage) tends to increase with time, and therefore indicates the

[20] This point will be discussed at length in Sects. 4.4 and 4.5 (cf. also Zimmerman, 2011).

[21] This must only be taken as a useful analogy. Strictly speaking, according to the theory of relativity, there simply are no times or spatial places at a fundamental level. It would thus be more accurate to say that "[…] all regions of spacetime are on a par, regardless of the particularities of their extension in spatial, temporal and null directions" (Pooley, 2013: 325).

[22] There are physical processes (e.g., neutral kaon decay) that are sensitive to the past-future orientation, but these processes are too "infrequent" and "exotic" to lead to strong conclusions (cf. Maudlin, 2007: 117).

[23] Albert (2000: 15) specifies that the only differences between S_F and its corresponding $S_{F'}$ have to do with where the magnetic fields are pointing. This specification is less innocent than it seems, since it leads Albert to claim that post-Newtonian theories are not – strictly speaking – time-reversal invariant, but that there remains in all of them a "curious vestige" of time reversal invariance. However, his view is highly controversial.

irreversibility of natural processes.[24] For instance, Lawrence Sklar claims that "[w]here there is no local entropic asymmetry, there is no future-past of time" (1992: 149). This claim will later be considered in greater detail, but I can already mention the main reason why it seems flawed: even if thermodynamics could yield the desired asymmetry, which is doubtful (cf. Uffink, 2001, Brown & Uffink, 2001, but also Poincaré's 'recurrence theorem'), this would not provide a *fundamental* explanation as to *why* the past appears to be fixed and the future open. Classical thermodynamics postulates both physical magnitudes – such as temperature, pressure, volume, entropy and heat – and laws stated in terms of these magnitudes, such as the Second Law. But these features are not believed to be *fundamental*. Many of them arise at a macroscopic level from the collective behavior of many microscopic entities. In general, quantum mechanics deals with the behavior of such microscopic entities, while its laws are – at least on the standard views – time-reversal invariant. So, even assuming that thermodynamics encodes a time asymmetry, it seems that this would at best postpone the problem: if there is no directedness in fundamental physics, then where does the thermodynamic asymmetry in time come from?

It might therefore be tempting for philosophers of physics to conclude that the asymmetry in openness between the future and the past is some sort of non-fundamental phenomenon, especially an epistemic phenomenon[25] (perhaps an artifact of the peculiar way our minds interact with the world). The main virtue of such a conclusion is a dialectical one: it explains away the awkward fact that the asymmetry has not yet been captured by our best physical theories and, therefore, preserves the reach of our understanding. This dialectical move is, by the way, pretty common in philosophy. There is indeed a fine tradition of dismissing awkward facts as non-fundamental features of reality. Kant ([1787] 1998: B51/A35), for example, argues that the Euclidean structure of space and time is not among the fundamental features of reality, but arises from the interaction of our sensory apparatus with the things in themselves. This idealist thesis has been explicitly developed to protect our knowledge of geometric truths from Hume's arguments that highlighted the fallibility of our epistemic devices. In brief, assuming that Euclidean geometry does not exist independently of us and, therefore, is not inferred from our ordinary inductive exploration of the world, it does not fall under Hume's skepticism.

Another example is given by the Everett, or 'many-worlds', interpretation of quantum mechanics. Some many-worlds theorists protect the possibility of some

---

[24] Entropy is a technical concept which allows for various definitions, but the general idea is easy to grasp: this is "[...] a measure of the extent to which the energy in a system has spread out in a disorderly fashion through the space available to it, and hence of how close a system is to its equilibrium state. If a system's energy has largely dissipated [...] it is in a high entropy state; if, by contrast, its energy is concentrated in just a few places, it is in a low entropy state" (Dainton, 2010: 47).

[25] Their conclusion may even be more radical. Sometimes the asymmetry between the future and the past is regarded merely as an illusion, i.e. as a perceptual phenomenon. This conclusion is, however, commonly denied, since the asymmetry has none of the marks of a regular illusion. In particular, it seems impossible to eradicate the asymmetry from experience in a way that would reveal its illusory character.

superpositions of systems at the macroscopic level by claiming that "[...] the most basic fact of laboratory experience – that experiments have unique outcomes – is an illusion" (Norton, 2010: 24). According to them, every time a quantum experiment with different possible outcomes is performed, all outcomes are obtained, each in a different world. For instance, many-worlds theorists affirm that Schrödinger's cat is both dead and alive, even before the box is opened. But since we do not see this macroscopic superposition – the cat is just dead, say, when we check – they conclude that there is another alive cat we cannot see, so that the definiteness of its death is an illusion. Here again, this theory requires us to dismiss some fact of experience as an illusion: we are actually deceived when we see just a dead cat. This dialectical move allows preserving the linearity of quantum mechanics (which states that objects can evolve into superpositions) without admitting that observation puts an end to this linearity.

However, it might seem that failing to capture particular phenomena, facts or entities is not sufficient grounds to doubt their fundamentality. For centuries physical models have not made any reference to quarks, but that did not prevent some ancient philosophers (e.g., Leucippus, Democritus, Epicurus) from rightly defending (supposing that it is right)[26] the view that the world is fundamentally composed of elementary particles.[27] Moreover, there are dozens of natural phenomena (e.g., northern lights, will-o'-the-wisps) that, although science had for long regarded them as subject-dependent, turn out to be objective features of reality. When impressed by the tremendous results obtained by physics through the last century, we get used to the idea that our best theories of space and time are telling us all that can be said about the nature of time, we just start to invert the reasoning. This is sometimes called 'the exclusivity dogma', namely the view that physics is an infallible guide to ontology, against which Yuval Dolev (2006: 189), for instance, warns us. If a phenomenon has all the marks of a fundamental and therefore non-epistemic one (like elementary particles in constitution processes), then it can legitimately be expected that physics must somehow characterize it. This leads to a principle, sometimes referred to as 'the principle of credulity' that seems reasonable to accept: intuitions must be preserved as long as science has not shown them to be wrong. Of course, this principle does not exclude that intuitions can be revised, but it states that they must be retained until they should be revised.

Moreover, assuming that the asymmetry is merely an epistemic phenomenon, the mechanism through which this phenomenon arises must be identified. There is a need to explain *why* – though the asymmetry has allegedly nothing to do with the nature of time – we have privileged epistemic access to the past rather than to the future. But such an explanation is rarely found in the literature. Most of the time, philosophers of science avoid the problem by claiming that potential solutions to

[26] The thesis that the world is fundamentally composed of elementary particles is controversial (see e.g., "wave function realism").

[27] Denying this point would amount to accepting a radical form of relativism 'à la B. Latour (2000)', who claims that Ramses II did not die of tuberculosis, since the Koch's bacillus was only discovered in the nineteenth century.

this issue are to be found outside their field of expertise. For example, Rudolph Carnap invokes here the role of psychology: "[...] all that occurs objectively can be described in science; on the one hand the temporal sequence of events is described in physics; and, on the other hand, the peculiarities of man's experiences with respect to time, including his different attitude towards past, present, and future, can be described and (in principle) explained in psychology" (1963: 37–38). In a similar vein, Paul Davies argues that "[t]he flow of time is an illusion. [...] And presumably the explanation for this illusion has to do with something up here (in your head) and is connected with memories [...]. So it's a feeling we have, but it's not a property of time itself [...]. Time doesn't flow. That's part of psychology" (cf. Dowker, 2020: 144). These kinds of buck-passing answers are, however, clearly unsatisfying. The issue raised by the origins of our intuitions regarding the nature of time cannot be skipped under the pretense that it could hypothetically be solved by psychologists (or other scientists). As, for instance, Mauro Dorato puts it: "[i]f a physical theory were in radical conflict with our experience of the world, and it could not give any explanation of the origin of such contrast, we should not invoke the illusoriness of our experience, but we would rather have good reasons to reformulate or even abandon the physical theory" (2008: 54).

Nonetheless, it must be acknowledged that there are projects that aim to explain the epistemic difference between the future and the past in certain physical facts (cf. Reichenbach [1956] 1971; Butterfield, 1984; Horwich, 1987; Craig 2001a, b; Callender, 2008). Apart from the details, all these projects involve the notion of causation, by appealing either to the *unidirectionality* of causation (causes occur prior to their effects), or to the *causal independence* of the past to the present (nothing that can now happen could have any effect on the past). However, although it is natural to associate temporal asymmetries with the idea that anything that can now happen can only have effects in the future (there is no backwards causation), there are reasons to doubt whether causation is the key of the mystery. For, unidirectional causation and causal independence appear to be neither necessary nor sufficient for fixity and openness. Consider the two following arguments.

First, reflection on permanentist thinking (always, everything always exists)[28] suggests that the causal independence of the past to the present is not *necessary* for the fixity of the past – or, to put it another way, the thesis that the future is fixed is compatible with the causal dependence of the future to the present. Consider the following case: by killing Archduke Franz Ferdinand of Austria, Gavrilo Princip (we may suppose) led to the outbreak of World War I. The latter event was causally dependent on the former. Yet, assuming that WWI has always existed (permanentism), it seems that one can still accept the causal claim that 'the outbreak of WWI is causally dependent on Gavrilo Princip's previous act of murder', without doubting that WWI could not have failed to take place. In other words, assuming that permanentism is true, one can question whether Gavrilo Princip had before him, at any time in his existence, a future that was 'open' rather than 'fixed'

[28] Cf. Williamson (2013: 25).

(without denying the causal role he played in the outbreak of WWI). Moreover, it seems that causal independence is not *sufficient* for fixity. A future that is a completely random continuation of the present is surely a future that is causally independent of the present. Yet such a 'random' future, far from being a 'fixed' future, might seem to be "[…] a paradigm of one type of openness (even if it represents a type of openness that brings with it no prospect of control over the course of events)" (Mackie 2014: 415).

Second, the unidirectionality of causation is not sufficient for openness. Consider the Gödel spacetime that admits closed time-like curves: "[…] if *P*, *Q* are any two points on a world line of matter, and *P* precedes *Q* on this line, there exists a time-like line connecting *P* and *Q* on which *Q* precedes *P*" (Gödel 1949a, b: 447).[29] In such a theoretical option, the future is clearly fixed (the sequence is closed and composed of a finite numbers of events), in spite of causation being unidirectional. The planting of a seed leads to the growing of a tree, which leads to the shading of a bench. Events causally related continue in the same way they would do in linear time. Of course, since the sequence of events is circular, one could theoretically plant a tree tomorrow in order to provide shade for a bench yesterday. But this is clearly not a case of backwards causation, since the shading at t_2 continues from the planting at t_4 in the forward direction, i.e. not through t_3, but through t_5, t_6, … t_1 (cf. Diekemper, 2005: 232). For these reasons, the question of causation appears to be largely independent of the question of temporal asymmetries and, therefore, fails to explain the asymmetry between what we can know about the past and what we can know about the future.

Finally, claiming that the asymmetry is merely an epistemic phenomenon betrays our intuitions. Although everybody agrees that we can know much more about the past than we can know about the future, *it cannot be the whole story*. We think of the open future and the fixed past in a much stronger sense. Unlike the spatial parts of which we have no memories and only few records (e.g., a distant planet, the center of the Earth), we do not think of the future as *out there*, waiting to be experienced. We rather think of it as *partially unsettled* until it has been made available to experience. This can be revealed by ordinary language: if I say that it is open whether my favorite football team (viz. Neuchâtel Xamax) will win the match tomorrow, I do not mean that the result is settled even though I don't know it yet. I rather mean that everything about tomorrow's match is still possible: perhaps my favorite team will win, or perhaps it will not. Our intuitions of openness seem thus to relate to worldly unsettledness regarding the future, rather than to our lack of epistemic access to what will happen. Of course, proponents of the epistemic approach might reply that the fact that their characterization is counterintuitive is of no consequence, since scientists have shown on many occasions that our intuitions (and especially our intuitions about time) are misleading. For example, it seems to us that 'our present' extends throughout the universe, while this intuition requires an objective notion of

[29] This possibility of closed timelike curves results from Gödel's exact solution of the Einstein Field Equations.

absolute simultaneity, which has been banned by Special Relativity (cf. Bourne, 2006). To that, two comments can be made.

First, all our intuitions are not equally important. Admittedly, our intuition of a common present has been denied by science (this will nonetheless be discussed in Sect. 4.6), but this intuition has come very late in human history. For centuries – as long as travel was on horseback, on foot, or in carriages – every village had its own peculiar time based on natural phenomena; there was then no reason to synchronize clocks between one place and another. It is only in the nineteenth century, with the development of the rail network, that the problem arose of properly synchronized clocks between different cities, and that the intuition of a universal 'now' emerged (cf. Rovelli, 2018: 47; Stephens, 1989). By contrast, the intuition of a fixed past and an open future has always been part of our manifest image of the world. At least, as long as we have viewed ourselves as agents capable of influencing the world in various ways, we have *presupposed* that the future was somehow open.[30] Therefore, it may not be as easy to deny the intuition of an open future than the intuition of a universal 'now', since these two intuitions play roles of varying levels of importance in our relationship to the world.

Second, an important consequence of the epistemic approach must be highlighted. If the openness of the future is merely to be explained by our great ignorance about what will happen, then, strictly speaking, the future is fixed. For example, the possibilities to act otherwise than how we actually act are not genuine possibilities, but merely *epistemic* ones. In such a picture (as in any other), there are two ways of seeing things: either humans can act freely, or they cannot. Assuming that human freedom should be preserved, the proponents of the epistemic approach have no choice but to accept a *compatibilist* theory (where the fixity of the future is not a threat to human freedom). As we will see in the next section, there are good reasons to believe that compatibilism is true, but – as compatibilists must themselves admit – even if this theory is true, it is not *trivially* true. Incompatibilists also provide powerful arguments to show that, when the future is fixed, no human act is free (precisely because, in such a picture, the possibilities to act otherwise are merely epistemic). So, since compatibilism is not trivially true (incompatibilists might be right, after all), there is a sense in which proponents of the epistemic approach endanger human freedom: they make it dependent on the non-trivial truth of compatibilism. By contrast, although substantial approaches to the open future (according to which the open future is not merely an epistemic phenomenon) do not ensure that some human acts are free, they do not make human freedom dependent on the non-trivial truth of compatibilism (nor on the non-trivial truth of incompatibilism). As a result, substantial approaches may seem more hospitable to human freedom than the epistemic approach. To be sure, this point is not against compatibilism; it merely highlights the fact that, assuming that human freedom should be

[30] This claim will nonetheless be qualified in the next section, when we will discuss the relationship between human actions and the open future.

preserved, proponents of the epistemic approach are forced to accept compatibilism, while this theory is, at best, non-trivially true.

Thus, although the asymmetry between what we can know about the past and what we can know about the future may, here again, end up being a *consequence* of the nature of time, it seems wrong to reduce the issue to its epistemic aspect. First, there are no good grounds for dismissing the asymmetry as an epistemic phenomenon (but mainly dialectical grounds). Second, there are no (or very few) satisfying attempts to identify the mechanism through which the epistemic asymmetry arises, if not grounded in the nature of time. And, finally, such an epistemic account betrays our basic intuitions that relate to the world itself (and not to the limits of our knowledge). For these reasons, it seems preferable to look for a more fundamental way of characterizing the asymmetry, which may ultimately explain *why* our knowledge of the future is not as vast as our knowledge of the past.

2.4 The Anthropocentric Attempt

Some philosophers tend to reduce the open future debate to the question of human abilities. They take the claim that 'the future is open while the past is fixed' to express the idea that 'humans can affect what will happen, but not what happened'. This move is natural, since our beliefs about our powers with respect to the future contrast sharply with our beliefs about our lack of power with respect to the past: whereas we do not deliberate about the past, our beliefs about opportunities, possibilities, alternatives, and so on, are all future-oriented. For instance, whereas there is no use crying over a broken window because once it has happened there is nothing we can do about it (except fixing it of course), we take to be (partially) within our power whether or not a future window is broken. Such an 'anthropocentric view' is, for example, defended by Stephan Torre who argues that "[t]he fact that so many [philosophers] have considered arguments against the open future by considering arguments that threaten our power over future events strongly suggests that [...] our notion of an asymmetry in openness between the past and the future is tied to an asymmetry in what we can affect or have power over" (2011: 361–362). In the same vein, Jenann Ismael claims that "[t]he future is as *open* as you are free to change your mind. And so understanding the sense in which the future is open really turns out to hinge on a proper understanding of the sense in which deliberation is unfixed by prior belief" (2016: 153). However, as we will see, it is one thing to say that the question of our abilities may inform us on the asymmetry between the open future and the fixed past; it is another to say that this asymmetry *is nothing but* an asymmetry in our abilities. This will become clear when we will consider senses in which the future and the past may respectively be said to be open and fixed in a world without humans (or before humanity emerged).

Another 'anthropocentric view' is endorsed by John Martin Fischer (1994, 2011) and Wesley Holliday (2012), who use the expression 'The Principle of the Fixity of the Past' to describe a thesis about a limitation on our abilities. More specifically,

Fischer (2011) focuses on the thesis, commonly referred to as 'theological fatalism', according to which God's foreknowledge and human freedom are *incompatible*: if there is a God who knows the entire future infallibly, then no human act is free. This thesis is generally motivated by the following line of reasoning: for any future act you will perform, if God infallibly believed in the past that the act would occur,[31] there is nothing you can do now about the fact that he believed what he believed, since (i) nobody has any control over past events (fixity of the past), and (ii) you cannot make God mistaken in his belief (infallible foreknowledge).[32] Therefore, "[…] there is nothing you can do now about the fact that he believed in a way that cannot be mistaken that you would do what you will do. But if so, you cannot do otherwise than what he believed you would do. And if you cannot do otherwise, you will not perform the act freely" (Zagzebski, 2017, cf. also Pike, 1965).

Although both incompatibilist and compatibilist replies can be made to this seemingly valid argument, Fischer's point is rather to reinforce it by clarifying premise (i), i.e. by providing an accurate characterization of the fixity of the past. As Fischer puts it: "[…] it is really unsatisfying simply to assert this [principle] as an *a priori* truth" (2011: 467); he therefore proposes to characterize the fixity of the past in terms of a limitation on our abilities, to the effect that we cannot do anything such that, were we to do it, the past would have been (or have had to be) different. According to Fischer, an agent cannot perform any action, the performance of which would require the past to have unfolded differently than it actually did. This characterization of (i) is meant to undermine any attempt to reject theological fatalism: if there is a God who knows the entire future infallibly, then you cannot do otherwise than what he believed you would do, since any different action would require a different past (in particular, any different action would require that God would have had a different belief), which is impossible. God's foreknowledge and human freedom are therefore incompatible.

Faced with this argument, compatibilists – who aim to reconcile God's foreknowledge with human freedom – must reject Fischer's characterization of the fixity of the past. In that respect, they might argue that, although nobody has an incredible power to *change* the past (i.e. to undo events that had already occurred in history), one has a more modest power, a power to do things at t such that certain events that actually occurred before t would never have occurred at all. For example, even if it is true that if I had visited my parents yesterday, the past would have (to have) been different (e.g., God would have had a different belief), I *could* have done

[31] A highly controversial presupposition that lies behind this claim is that God's belief may be a past event, while this seems to threaten God's immutability (i.e. God is unchanging in his attributes).

[32] It is worth noting that, in what is commonly presented as the "classical" formulation of the argument, Nelson Pike (1965) restricts the fixity of the past to a specific class of facts about the past: *the hard facts* (as opposed to *the soft facts*). Roughly, whereas hard facts are "fully accomplished", "over-and-done-with" and so forth in the past (e.g., Caesar died on the steps of the Senate), soft facts are not (e.g., Caesar died 2064 years before I wrote these lines). However, as Fischer makes clear, this distinction between hard and soft facts does not match with the distinction between what is out of any agent's ability to affect and what is not (cf. Fischer 2011: 466).

so, provided I was under no coercion or compulsion to not visit my parents. Of course, in such a case, different events would have occurred instead, but no event would have both occurred and then been undone by my action. Thus, once one realizes that compatibilism does not involve any commitment to a power of undoing the past, it seems that Fischer's characterization of (i) fails to undermine any attempt to reject theological fatalism. From a compatibilist point of view, what is right about the Principle of the Fixity of the Past is that we cannot *undo the past*; what is wrong about Fischer's characterization of this principle is that it goes further, stating that we cannot do anything that *requires a different past*. In short, compatibilists believe that one *can* perform an action that is inconsistent with the actual past (and so Fischer is wrong), even though one *will not*.[33]

The question as to whether God's foreknowledge can be reconciled with human freedom is somehow similar to a more general and well-known question in the history of philosophy, 'the free will problem', which concerns a disputed incompatibility between *free will* (i.e. the ability of persons to make decisions of the sort for which one can be morally responsible) and *determinism* (i.e. the thesis that at any time the world has exactly one possible future, given the past and the laws of nature).[34] Unfortunately, despite all of the work philosophers have devoted to it, there is no single specification of the free will problem. Part of the reason is that a completely neutral formulation of the problem can hardly be found. So, although I doubt that what follows will meet general approval, here is how Robert Kane introduces the problem:

> […] suppose Jane has just graduated from law school and she has a choice between joining a law firm in Chicago or a different firm in New York. If Jane believes her choice is a free choice (made "of her own free will"), she must believe both options are "open" to her while she is deliberating. She could choose either one. (If she did not believe this, what would be the point of deliberating?) But that means she believes there is more than one possible path into the future available to her and it is "up to her" which of these paths will be taken. [However] if determinism is true, it seems there would not be more than one possible path into the future available to Jane, but only one. It would not be "up to her" what she chose from an array of alternative possibilities, since only one alternative would be possible (2007: 6).

One reason to complain about Kane's formulation of the 'free will problem' is its use of the locution "it is up to her". For, both incompatibilists and compatibilists agree that human actions are not *epiphenomenal*. Therefore, even if determinism is

[33] This kind of compatibilism is often called 'backtracking compatibilism' (cf. Holliday, 2012). Although this position seems to be the best compatibilist reply that can be provided against Fischer's argument, it is not the only possible one. For example, one might adopt a 'miracle compatibilism' according to which if I had visited my parents at *t*, then all of history would have been the same until shortly before *t*, at which time a violation of the actual laws of nature (so-called 'divergence miracle') would have allowed me to visit my parents.

[34] See Fischer (2014) for a discussion of the parallels between the argument for incompatibilism about God's fore-knowledge and human freedom, and the argument for incompatibilism about determinism and human freedom.

true, compatibilists will claim that "it is up to Jane" whether she will work in Chicago or New York (at least as long as her choice is made for her own reasons).

As in the previous debate, there is a long-standing tradition of dividing up the conceptual terrain in two families of positions. Traditionally, *incompatibilists* are those who think that free will is incompatible with the world being deterministic, while *compatibilists* reject this thought. To return to Kane's example, an incompatibilist will typically claim that, if determinism is true, Jane has no ability to choose how her own future will unfold (she has no control over the past or the laws of nature), while a compatibilist will argue that, although there might be only one possible way the future might unfold, Jane still has the ability to choose to work in Chicago or New York (she is under no coercion or compulsion, after all). It is worth noting that neither incompatibilists nor compatibilists are committed to the further claim that Jane does, in fact, have free will. However, as Michael McKenna and Justin Coates make clear: "[…] many compatibilists (but by no means all) do think that we are sometimes free. And though some incompatibilists remain agnostic as to whether persons have free will, most take a further stand regarding the reality or unreality of free will" (2015: §1.4). In the philosophical literature, *libertarians* are the incompatibilists who argue that at least some persons have free will (and, therefore, that determinism is false), while *hard incompatibilists* (or *hard determinists*) have a less optimistic view, holding that determinism is true and, therefore, that no persons have free will.

From a contemporary perspective, the conflict between incompatibilists and compatibilists lies, at least partially, in a disagreement over the meaning of 'can' (or 'have the power') and related expressions, such as 'could have done otherwise' (which does not imply that the conflict is merely verbal). According to incompatibilists, we are able to do otherwise only if our doing otherwise is *a possible continuation of the past consistent with the laws*. It thus appears that, if determinism is true, there is only one possible continuation of the past consistent with the laws, and so no human action is free. As Kadri Vihvelin puts it: "[w]hat we actually do is the only thing we are able to do" (2003: §5). By contrast, compatibilists insist that 'can' (or 'have the power') should be understood as a counterfactual expression: when someone says 'you can (or have the power) to do something', it simply means that 'if you want (or try) to do it, you shall do it'. For example, saying that 'you can jump over this fence' means that 'you will jump over it, if you want to or try to'. It is therefore wrong to claim that, if determinism is true, no human action is free, since our actions counterfactually depend on our choices, which in turn depend on the reasons we take ourselves to have (at least in the normal case, where there is no coercion or compulsion). If our reasons were different, we would choose otherwise, and if we chose otherwise, we would do otherwise. And, it seems that our reasons *can* be different, at least in the sense that we, unlike animals or young children, have (i) the ability to *critically evaluate* our reasons (beliefs, desires, values, principles, etc.) and (ii) the ability to *change* them (this is sometimes called the 'can' of freedom and choice).

Of course, this is not the place to settle these debates, even if I am inclined to think, as Fischer does, that the compatibilist interpretation of 'can' (or 'have the power to') is contextually inappropriate. After all, it is uncontentious that I have the general ability and 'know-how' to refrain from visiting my parents (even in the scenario in which God has foreknowledge that I will in fact not visit them), just as it is uncontentious that Jane has the general ability and 'know-how' to accept a position in Chicago or New York (even in a scenario in which determinism is true). But this does not seem to be the conception of ability in question. The whole point of the human freedom debate is to question whether, in some specific contexts (e.g., there is an omniscient God, determinism is true), we have the ability to do otherwise with respect to ordinary actions. Having a general capacity (relevant skills and know-how) is *not* sufficient for the ability in question, since one might have the general capacity while being blocked from exercising it in various ways! So, it does not seem 'dialectically kosher' simply to assume, in these two examples, that Jane and I have the ability to do so and so. As Fischer puts it: "[...] one cannot simply import ordinary views about our powers into the philosophical context of an evaluation of the argument for the incompatibility of God's foreknowledge [or determinism] and human freedom – a skeptical argument that explicitly challenges these ordinary views about powers" (2011: 471).[35]

Anyway, despite what Fischer suggests, what incompatibilists and compatibilists seem to agree upon is that nobody can perform an action that changes the past (which does not imply, according to compatibilists, that nobody can perform an action that is inconsistent with the past), not even time-travelers, contrary to what many Hollywood movies suggest. In a model with a single past, changing the past clearly involves contradictions: e.g., the time-traveler bets on the victory of his favorite team in 1976, and does not bet on the victory of his favorite team in 1976. It is not as if there were two versions of the past: the original one, and a second version with the time-traveler playing a role. There is only one past and two perspectives on it: the perspective of the younger self, and the perspective of the older time-traveler. If these perspectives are inconsistent (e.g., if an event occurs in one but not in the other), then the time-travel scenario is incoherent.[36] However, although time-travelers cannot make the past different from the way it was, this does not mean that they must be entirely powerless in the past: they *can* participate in it, in particular they *can* do anything that happened. For example, if tomorrow Sam travels back to 1976, then the past already contains Sam appearing out of nowhere in 1976, as well as it contains all the actions Sam performed there and all the consequences of his actions. Sam will be causally effective in the past, but he will not

[35] My own reply to theological fatalism will be developed in the next chapter (Sect. 3.5). The basic idea is that this view rests on a confusion between two different notions: to be true, and to be inevitably true.

[36] For an exception, see Loss (2015) who argues that time-travelers can change the past even if time is linear. In that respect, Loss introduces a sort of branching model where all the branches are ordered in a linear way.

bring about any change: before he travels what *will* happen to him *happened* in the past (cf. Andreoletti, 2020 and Lewis, 1976).[37]

Now the central question is the following: is it a good suggestion to reduce the open future debate to the question of human abilities? Although I must acknowledge that our power and lack of power over certain events is constitutive of our pre-theoretic understanding of openness and fixity (cf. Torre, 2011: 361), I think that it is a bad suggestion. First, unless we find some common ground between incompatibilists and compatibilists on what the expression 'can' (or 'have the power') means (which seems hopeless – the free will debate has been raging since Plato and Aristotle, cf. Irwin, 1992), the risk is to privilege one or the other of these positions. For example, supposing that there is an omniscient God or that determinism is true, *can* humans choose how their own future will unfold? As has been shown, the answer depends on whether we adopt an incompatibilist or a compatibilist point of view: if 'can' means that humans have the freedom to do otherwise, incompatibilists immediately answer 'no', while at least some compatibilists might answer 'yes' (there is at least one counterfactual conception of 'can' according to which humans can choose how their own future will unfold). Therefore, since we do not want our characterization of the temporal asymmetry to presuppose the truth (or the falsity) of incompatibilism or compatibilism – two venerable positions in the history of philosophy – it seems preferable to look for another option. It is worth noting that, although the failure of determinism will be introduced in the next section as a *necessary* condition for the future being open, this does not preclude compatibilism, since this does not preclude the possibility for humans to act freely when the future is fixed; in such a case, humans remain causally efficient, after all.

Second, reducing the open future debate to the question of human abilities is objectionably agent-centered. It might indeed seem that the asymmetry between the open future and the fixed past was prior to the existence of any agent. For example, it might be argued that, 100 million years ago, it was open whether dinosaurs would disappear and humanity would emerge. Moreover, there is at least one sense in which the future may be said to be open that does not involve any agent: time could come to an end, with no ontological commitment to future things standing in the way (cf. Correia & Rosenkranz, 2018: 99). It indeed seems possible that the future is open not simply in terms of *how* it will unfold, but also in terms of *whether* it will unfold (this will be discussed at length in Sects. 2.7, 2.8 and 2.9). Taking this possibility seriously (as a number of physicists and theologians do), while it obviously exceeds anything humans may claim to have power on (no matter what we mean by 'power' here), it must be concluded that the question of the open future should not be reduced to the question of our abilities. After all, there is nothing strange in supposing that there are senses in which the future may be said to be open in a world

[37] The situation might be different when it comes to the branching-tree conception of time: at t_1 Sam can travel to an earlier time t^* to prevent an event e to occur, provided that (i) t^* is located on a different branch from the one he departed and (ii) e does not occur on t^*. However, whether this case should count (or not) as a genuine case of past alteration is debatable.

without humans (especially if determinism is false and there being no fact about what will happen, cf. Sect. 2.9).

The point of this section was not to minimize the importance of the human freedom debate (even if determinism and God's foreknowledge are no longer at the heart of our highest concerns – most physicists think that the world is not deterministic (at least with respect to quantum mechanics) and theological questions have progressively been banned from scientific inquiry). This debate is crucial, especially with regard to moral responsibility, because it is generally agreed that having free will is a necessary condition of being morally responsible, so that if determinism precludes free will, it also precludes moral responsibility. The idea was rather to show that, although a reflection on our abilities (what we can or cannot do, what 'can' truly means, etc.) contributes to the understanding of the openness and fixity phenomena, the asymmetry between the open future and the fixed past should not be reduced to an asymmetry of mutability. As Penelope Mackie puts it: "[...] Fischer's principle seems more appropriately regarded as a (controversial) claim about a *consequence* of the fixity of the past, rather than an expression of what it means to say that the past is fixed" (Mackie, 2014: 414). One should therefore look for a less anthropocentric characterization of the temporal asymmetry, and ultimately assess whether it is compatible with human freedom.

2.5 Physical Indeterminism

Suppose that the world is *physically indeterministic* in the sense that its future history is not nomologically necessitated by its current history: "[t]o say that the future is open might only be to say that the future is not nomologically determined in this sense" (Pooley, 2013: 337). In other words, the way the world is, up to a certain time *t*, together with the laws of nature, does not necessitate the way it is at any future time *t**. Of course, there can only be a unique and actual way the world is at *t**, but, assuming *physical indeterminism*, this way "[...] need not be the only one compatible with the actual laws and the way the world is up to and including *t*" (Pooley *id*). According to this understanding, the future is open at a certain time *t* only if, given how the world is up to *t* and what laws obtain, there are several possible ways the world might be at some future time *t**.[38]

There are at least three scenarios in which indeterminism might be the case. First, following Nancy Cartwright (2008), it could be that there are not any laws of nature, such that the world is a disordered jumbled place of random facts. Second, it could be that there are laws of nature, but they do not cover all types of situations that may arise in our universe. For example, perhaps what we call 'laws of nature' are local laws that only describe what happens in the part of the universe we can

[38] Cf. Earman, 1992 for a detailed presentation of the different kinds of determinism and indeterminism that can be found in the philosophy of science. See also Prior, 1967, Honderich, 1988, Van Inwagen, 1989.

observe, but there are other parts of the universe for which there are not even local laws that apply. Third, and this is what many physicists think follows from quantum mechanics, it could be that there are laws of nature that govern what happens in all types of situations, but these laws do not describe what will happen as a matter of necessity; they only describe what will happen as a matter of chance. Specifically, although the central dynamical law of quantum mechanics – Schrödinger's equation – is purely deterministic (it does not make any reference to chances), many physicists think that there must be some additional laws describing the chances that there will be a collapse of the wave function (cf. Albert, 1994: 80). If this is true, i.e. if there must be probabilistic laws governing the wave function, then the laws of the universe will turn out to be indeterministic.

The claim that the 'open future' view *is nothing but* the view that 'there are various futures that can nomologically issue from the current history' sounds plausible. A reason is that *physical determinism*, which is correspondingly understood as the negation of *physical indeterminism* (once you have fixed the world history up to a certain time *t* and the laws of nature, you have fixed the world history at any future time *t**), seems to commit to the fixity of the future. After all, if *physical determinism* is true and, therefore, if there is only one possible way the world history can unfold, then it is unclear how the future could still be called open. Specifically, if it is *necessary*, given the facts about how things are up to *t* and what laws obtain, that the world will be a certain way at *t**, then it seems *settled* that it will be that way, which is just to say that the future is *fully settled* at *t*, and hence that the future is fixed. This point may, in particular, be illustrated by Laplace's intelligence (or demon as it is often called). Assuming determinism, if a demon (which should be construed as a 'super-scientist', cf. Popper 1988: 34) knows the precise location and momentum (i.e. the product of the mass and velocity) of every particle of the universe, i.e. has an access to the complete specification of past and present world-states, he can predict their future locations and momenta from the laws of nature. Although Laplace seems to have another kind of determinism in mind (expressed in terms of predictability) and only the laws of classical mechanics, his thought experiment tends to show that if determinism is true, then the future is predictable and hence fully fixed (cf. Laplace [1814] 1951).

John Mackie, for example, endorses such a physical characterization of the open future when he claims that the laws of nature[39] cover the ways in which things become fixed or remain open. As he puts it, "[...] the universe needs to know where to go next" and the laws of nature cater to this need (1974: 225). For instance, if two electrons collide, then the laws of quantum mechanics dictate what the future possibilities are: they fix that whereas none of these electrons will turn into a proton or into a water molecule, there is a probability that these electrons will deflect at this or that angle. According to Mackie's view, the current world history and the laws of nature are thus together both *necessary* and *sufficient* to establish whether the future is open or not. If there is only one way the future can nomologically issue from the

[39] Mackie speaks rather of the "laws of working" (cf. 1974: 225).

current world history, then the future is fixed (*physical determinism*) and, conversely, if there is more than one way the future can nomologically issue from the current world history, then the future is open (*physical indeterminism*).[40]

However, physical indeterminism is often deemed insufficient for characterizing the openness of the future. For example, Peter Geach argues that if there is a current world history from which the future can nomologically issue only in one way, then the future is fixed, *but the converse does not hold*. As he metaphorically puts it, considering a book, "[...] even if the text of later pages is not determined by the text of earlier pages, there may nevertheless be a completely fixed text on those pages which we have not yet turned over" (1973: 208). Although the relevance of this metaphor might be questioned, the idea it expresses is quite clear: the future may be fixed without being nomologically necessitated by the current world history. In particular, it appears that in a permanentist picture of the world (where always, what exists always exists), the future might be regarded as fixed in spite of the world being indeterministic with regard to what will happen. As Jiri Benovsky puts it: "[...] if the inventory of all there exists in the universe includes all past and future times and entities as well as present ones, then the future is, metaphorically speaking, 'already there' (as well as the past and the present). [...] No surprise, then, that under such a view the future is already fixed, in a metaphysical way" (2013: 160). For example, it seems reasonable to think that the existence of my grand-son (which is not located in the current world history) fully settles that I will have children in the future, no matter whether the world is indeterministic (or not), at least assuming that Kripkean necessity of origin holds.[41]

More controversially, it has been argued that *physical indeterminism* is not even *necessary* for the future being open. After all, assuming *physical determinism*, what is necessary is not, strictly speaking, the way the world is at any future time t^* (nor the way it has been up to t), but rather the conditional that *if* the world has been a certain way up to a certain time t and has certain laws L, *then* it will be a certain way at any future time t^*. Both the antecedent and the consequent of the conditional are contingent in this picture. It might then appear that all that we are entitled to say is that "[...] the consequent of the conditional is settled if the antecedent is; if there is any unsettledness in the antecedent, this may bleed over into the consequent" (Barnes & Cameron, 2009: 300). Following this line of reasoning, Barnes and Cameron conclude that it is wrong to claim that *physical determinism* commits to the fixity of the future, since all that this thesis entails is that the future is only open if there is unsettledness in the antecedent of the conditional (i.e. either in the world history up to t, or in the laws of nature which obtain). If both the laws and the history are settled, then the future is fixed, but if either the laws or the history is unsettled,

---

[40] As a reminder, *physical indeterminism* does not necessarily entail complete lawlessness but may be expressed by means of probabilistic laws.

[41] However, as we will see (Sect. 2.7), there are interesting attempts to reconcile particular versions of permanentism with the open future (cf. Barnes & Cameron, 2009, 2011; Cameron, 2015 and Skow, 2015).

then the future might be open. In other words, the future could be called open even if indeterminism is false.

However, as suggested above, this latter argument is highly controversial. For example, Sven Rosenkranz (2013) points out that, by Barnes and Cameron's own view, the sense in which the current world history might be unsettled is not the sense in which the future is said to be unsettled by being open. In particular, the current world history might be said to be unsettled in the sense that, prior to opening the box, Schrödinger's cat is neither determinately alive nor dead. But this is clearly not a case of the world being unsettled about which range of completely determinate options obtains, as the openness of the future is meant to be.[42] According to Calosi and Wilson (2019), for example, the superposition of states (as a source of quantum metaphysical indeterminacy), must rather be seen as involving a state of affairs whose constitutive entity (Schrödinger's cat) has a determinable property (*having a certain life status*), but no unique determinate of that determinable (*being alive* and *being dead*). Therefore, it is hard to make sense of the claim that the unsettledness in the antecedent can "bleed" into the consequent, since Barnes and Cameron are dealing with at least two different kinds of unsettledness: one that concerns the current world history and the laws of nature (and that may include quantum metaphysical indeterminacy), the other that concerns the plurality of ways the future might develop. To be sure, the claim is not that it is impossible to provide a unified account of both quantum and future unsettledness, but only that if the openness of the future is defined as Barnes and Cameron do, i.e. as the world's being unsettled about what range of completely determinate options obtains, then it radically differs (by their own view) from the unsettledness that lies in the superposition of quantum states.

In a nutshell, the first reason why *physical indeterminism* does not appear to be appropriate for characterizing the asymmetry in openness between the future and the past is that, although this thesis seems to be *necessary* for the future being open (the future is intuitively fixed in a fully deterministic world), it does not seem to be *sufficient*. Since *physical indeterminism* is available to permanentists who can clearly deny that the future is genuinely open (although they might not be forced to), it appears that *something more* than *physical indeterminism* is required to fully characterize the asymmetric nature of time. That is all the more evident as there are interesting senses in which the future may be said to be open that are not constrained by the current world history and the laws of nature (at least assuming that the world is not fully deterministic). After all, the facts about how the future might unfold need not be fully grounded in physical facts. Perhaps the only sense in which the future turns out to be open is that time will come to an end within 1 s. This logical possibility should not be ruled out on the grounds that it does not exclusively rest on a naturalistic assumption, or so it could be argued (cf. Barnes & Cameron, 2011: §3).

A second, and perhaps more fundamental, reason why *physical indeterminism* (alone) does not appear to be appropriate for accommodating the asymmetry has to do with the symmetric nature of physical laws. As a reminder, for every physically

[42] Barnes and Cameron's account for the openness of the future will be detailed in Sect. 2.7.

allowable sequence of states, the inverse sequence of time-reversed states is also physically allowable (cf. Sect. 2.3). Roughly speaking, given that the laws of nature are time-reversal invariant, it seems that the world cannot be considered as physically indeterministic with regard to the future without being considered as physically indeterministic with regard to the past.[43] It would indeed be arbitrary to claim that the current world history up to a certain time *t*, together with the time-reversal invariant laws of nature, is compatible with various alternative futures, but not with various alternative pasts. Hence, any good argument from *physical indeterminism* in support of the open future seems to be, at the same time, an equally good argument with regard to the open past (cf. Markosian, 1995). Therefore, since *physical indeterminism* entails that if the future is open, so must be the past; it clearly fails to accommodate any asymmetry in openness between the future and the past. For these reasons, it seems preferable to look for another notion that may ultimately be combined with *physical indeterminism* in order to provide an accurate characterization of the asymmetry.

2.6 Counterfactual Dependence

The most famous example of an account using modal resources in order to characterize the asymmetry between the open future and the fixed past has been provided by David K. Lewis:

> I suggest that the mysterious asymmetry between open future and fixed past is nothing else than the asymmetry of counterfactual dependence. The forking paths into the future – the actual one and all the rest – are the many alternative futures that would come about under various counterfactual suppositions about the present. The one actual, fixed past is the one past that would remain actual under this same range of suppositions (1979a: 462).[44]

Basically, "[...] the future is open *because* if the present were different the future would be different, whereas the past is fixed *because* if the present were different the past would remain as it in fact is" (Barnes & Cameron, 2011: 7). For example, suppose that I were watching a movie instead of writing these lines. Then clearly tomorrow would be different also; e.g., I would fall behind in writing this section and presumably feel guilty about this. By contrast, it might seem that there is no true

[43] Cf. Geach, 1973; McCall, 1976; Lewis, 1979a; Markosian, 1995; Torre, 2011; Barnes & Cameron, 2009. However, this can be disputed: whereas the time-reversal invariant laws are deterministic, the non-deterministic laws (e.g., Albert wave function collapse law) need not be time-reversal invariant.

[44] It is worth noting that, although Lewis often writes as if his task were to explain *the fact* that the past is fixed and the future open, his principal explanandum is rather *our intuition* that it is so. In that sense, it is doubtful that, for Lewis, our intuition of an open future and a fixed past has an objective correlate, so that his account should probably be classified as an epistemic account. However, it is not uninteresting to take Lewis's account more seriously, i.e. as if it were saying something about how time truly is.

counterfactual about how the past would be different if the present were somehow different. Intuitively, if I were watching a movie instead of writing these lines, what I wrote yesterday would not be affected. As Lewis puts it, "[i]t is at best doubtful whether the past depends counterfactually on the present, whether the present depends on the future, and in general whether the way things are earlier depends on the way things will be later" (1979a: 455).

Of course, counterfactuals are context-sensitive. Given that present conditions have their past causes, one can, as Lewis acknowledges – in *atypical contexts* – set things up to accept as true that, had the present been different, the past would have been as well. This is commonly called a *back-tracking argument*. For example, suppose that Jim and Jack quarreled yesterday, and that Jim is a prideful fellow who would never ask someone he quarreled with the day before for help (cf. Downing, 1959). One can easily accept the truth of ⌜If Jim asked Jack for help today, there would have been no quarrel between the two men yesterday⌝. It thus seems that whether or not an assertion of a counterfactual claim is true depends on what facts do matter in the context of assertion. After all, there is no absolute evaluation of counterfactual claims; the evaluation is always relative to a preselection in the kind of facts that serve as points of reference. Therefore, in order to invalidate back-tracking arguments, and hence yield the desired temporal asymmetry of counterfactual dependence, proponents of the modal approach must explain what facts do matter for the evaluation of counterfactual claims.

In that respect, Lewis (1973a, b, 1979a) provides (i) an analysis of counterfactuals based on comparative similarity of possible worlds and (ii) an account of what kind of facts do matter when comparing possible worlds. Basically, his idea is to take what is open to be what could happen under various counterfactual assumptions concerning the present *given* a metric of similarity for possible worlds. The evaluation of counterfactuals thus requires the comparison of worlds for similarity to the actual world. According to Lewis's analysis, "[…] a counterfactual 'If it were that *A*, then it would be that *C*' is (non-vacuously) true if and only if some (accessible) world where *A* and *C* are true is more similar to our actual world, overall, than is any world where *A* is true but *C* is false" (1979a: 465). For instance, the counterfactual ⌜If Jim swallowed his pride, he would ask Jack for help⌝ is true iff every world that makes the antecedent true (Jim swallowed his pride) *without a great divergence with actuality* is a world that also makes the consequent true (Jim would ask Jack for help).

Then, given that similarities and differences among worlds are not all equal (some of them have more 'weight' than others), Lewis provides further rules of how to order worlds for similarity. Specifically, he states that *physical facts* (e.g., the laws of nature) are the facts that matter when comparing worlds. In general, the less a world violates the laws of nature, the more similar it is to our world. Arguably, this account invalidates back-tracking arguments: the counterfactual ⌜If Jim asked Jack for help today, there would have been no quarrel yesterday⌝ is usually false *because* the consequences of supposing that there was no quarrel would make *more difference* (from a physical point of view) to the world than those of supposing that somehow Jim swallowed his pride. In Lewis's terms, the world (or, more precisely, the type of worlds) in which Jim somehow swallowed his pride is closer to ours (physical laws are not violated) than the world in which there was no quarrel. Accordingly, if the present were different the past would be the same (e.g., if Jim asked Jack for help today, the quarrel would nevertheless have taken place yesterday), but the same causes would somehow fail to cause the same present effect (e.g., Jim would somehow have swallowed his pride and asked Jack for help despite yesterday's quarrel).

At first sight, Lewis's account does display the desired temporal asymmetry of counterfactual dependence, but a closer look reveals important difficulties. First, on Lewis's view, the facts about what is open depend on the physical facts that determine the similarity metric for possible worlds, while it has been shown that such naturalistic resources cannot do the job. In particular, it is hard to see how Lewis could get an asymmetry in counterfactuals out of physical laws that are time-reversal invariant. At the end of his paper, Lewis suggests that thermodynamics might play a crucial role in his metric: the Second Law, when extended from isolated systems to the whole universe, indicates that a given event tends to have an isolated cause but many spread-out effects (it leaves an enormous multitude of traces into the future). As a result, "[…] only a small miracle is needed to allow an event that did happen not to have happened, keeping the past of that event fixed. But a gigantic one would be needed to allow that event to have happened keeping the future exactly the same" (cf. Sklar, 1993: 403). It can therefore be supposed that the more a possible world avoids 'big, widespread, diverse' violations of the Second Law (i.e. the more it preserves the entropic spreading situations), the more it is similar to our world.

However, this Reichenbach-inspired strategy, which has already been encountered in Sect. 2.3, seems to fail. First, as John Earman (1974) points out, applying the thermodynamic concept to systems as diverse as decks of playing cards, human beings, printed records, and so on, involves a considerable extension of the concept beyond its established range of legitimacy, and there is no guarantee that the extension is legitimate. Second, if Harvey Brown and Jos Uffink (2001) are correct, then the Second Law of thermodynamics does not presuppose any time asymmetry at all; it is simply a relation between a few variables at equilibrium. And there is no question that thermodynamics, if not its Second Law, makes time-asymmetric claims (cf. Callender, 2001). Henry Poincaré reaches a similar conclusion when he introduces his 'recurrence theorem' (against Boltzmann's view), according to which the particles in any energetically isolated state will, given enough time, return to a state that is arbitrarily analogous to their initial state: the entropic asymmetry may break. Finally, even if thermodynamics could yield the desired asymmetry, this would not provide a *fundamental* explanation as to why the past appears to be fixed and the future open. As has previously been argued (cf. Sect. 2.3), thermodynamics is a 'phenomenal' science; that means that its variables range over macroscopic magnitudes, such as temperature, pressure, volume, heat, and entropy. These magnitudes arise from the collective behavior of many microscopic entities, which is governed by the time-reversal invariant laws of quantum mechanics. Therefore, claiming that the asymmetry in time is to be grounded in the increase of entropy (i.e. in the spread-out effects of events) simply postpones the problem: if there is no directedness in fundamental physics, what then grounds the entropic asymmetry in time?[45]

Of course, as Barnes and Cameron suggest, one can in principle "[...] agree with Lewis that openness is to be analyzed in terms of counterfactual dependence on the present and disagree with his account of what facts get fixed for the evaluation of counterfactuals" (2011: 7). But this option leads to a *dilemma*. Either there is something special about the facts that matter (that get held fixed) in the evaluation of counterfactuals, or there is not. If there is, then extra non-naturalistic resources are to be found in order to say what is special about these facts; we are back to square one. If there is not, then Lewis's attempt is completely irrelevant to the explanatory project: "[t]here needs to be something special about the facts to be held fixed that explains *why* holding them fixed is relevant to openness" (2011: 8). After all, supposing that some facts are *arbitrarily* held fixed, *why* should it matter what would have happened *given* these facts rather than others?

Penelope Mackie (2014) also provides several objections against Lewis's account, and especially against his characterization of the fixity of the past in terms of counterfactual dependence. In particular, she argues that there is an ambiguity in the claim that we *usually* take the past to be counterfactually independent of the present. Let me explain. Lewis explicitly endorses the thesis that backward counterfactuals are, as well as forward counterfactuals, *usually* assessed in a way that keeps the past (relative to the time of the antecedent) fixed (cf. Lewis, 1979a: 458).[46] For example, when we assess the backward counterfactual ⌜If Jane had handed in her essay on Monday, she would have revised it properly first⌝, we assume that the past (relative to the antecedent, i.e. the past until Monday) in the counterfactual situation is exactly the same as the actual past. That is why Lewis takes his conclusion that there is no true backward counterfactual (unless the consequent is true, i.e. unless Jane's essay has indeed been revised properly) to be in line with usual practice.

However, as Penelope Mackie points out, this is "[...] plainly false as a description of the way in which we *usually* evaluate counterfactuals" (2014: 403).[47] There are dozens of examples in which we naturally suppose that if the situation had been different at a time *t*, then some features of the past relative to *t* would have been different also, otherwise the situation would not have been thus. Downing's case involving prideful Jim and angry Jack may obviously serve as a good example, but Mackie imagines many different cases: (i) ⌜If there had been ice on the pond this morning, the temperature last night would have been lower than it actually was⌝; (ii)

[45] For an extended discussion of this topic, see Albert, 2000, Loewer, 2007, Wallace, 2013, Ismael, 2016, chap. 6.

[46] 'Backward counterfactual' refers to one whose consequent is about an *earlier* time than any that its antecedent is about. In this sense, 'backward counterfactual' need not say that the earlier would have been different had the later been different and must, therefore, be distinguished from the counterfactual involved in a back-tracking argument. By contrast, 'forward counterfactual' refers to one whose consequent is about a *later* time than any that its antecedent is about (cf. Bennett, 1984: 57).

[47] Bennett (1984: 77) develops a similar objection to Lewis's account.

'If the roof had been intact today, it would not have been hit by a falling tree yesterday'; etc. In this way, Mackie shows that back-tracking cases (which are a kind of backward cases) are much more common than Lewis claims and, therefore, that they do not require atypical contexts.

As we have seen, Lewis resolves backward cases by keeping the past relative to the time of the antecedent fixed, with the result that all backward counterfactuals (including back-tracking ones) usually come out false. But this solution is only one among others and, as Mackie puts it, Lewis has no right to "[...] call this (*non-back-tracking*) resolution 'the standard resolution'" (2014: 404), especially considering that it outlaws back-tracking cases which are, as indicated above, very common. According to Mackie, Lewis has taken a feature (the 'keeping fixed' of the past relative to the time of the antecedent) that may with some plausibility be regarded as a feature of the resolution of forward cases, and has extended it, quite implausibly, to the resolution of backward cases. Therefore, since Lewis's resolution of backward cases is insufficient – sometimes we keep the past fixed, but sometimes we do not (especially in back-tracking cases) – it appears to be incapable of yielding a satisfactory characterization of the fixity of the past.

It can therefore be concluded that (i) Lewis cannot get an asymmetry in counterfactuals out of only physical facts, and (ii) even if he could, his account would fail to describe the way we *usually* evaluate backward counterfactuals. There are many examples in which we naturally suppose that if the present had been different, then the past would have been different also (cf. back-tracking cases). As a consequence, Lewis's account fails to display a temporal asymmetry of counterfactual dependence that accounts for the asymmetry in openness between the future and the past, and must as such be rejected. Of course, there is no certainty that these criticisms of Lewis's account can be extended to any modal account. After all, there might be some modification of Lewis's attempt that would escape these criticisms – Mackie expresses serious doubts about this (cf. 2014: 413). One thing is certain: contrary to what Barnes and Cameron (2011) suggest, it is not enough to supplement Lewis's counterfactual analysis with the right (non-naturalistic) ideology to ground the desired asymmetry between what is open and what is fixed. The modification has indeed to be more radical, insofar as the past *does* sometimes counterfactually depend on the present.

2.7 Metaphysical Indeterminacy

A position that is in the minority, perhaps, but deserves ever greater attention is the claim that the fixity of the past and the openness of the future must be characterized in terms of metaphysical determinacy and indeterminacy, respectively. The main idea is that whereas it is *fully determinate* how the past did turn out, it is *partially indeterminate* how the future will turn out. The way in which how the future will turn out fails to be *fully determinate* is to be understood as a brute fact: sometimes it is simply unsettled how the world is at its most fundamental level. At this point, it is worth noting that among the various phenomena that may be thought of as types of worldly unsettledness, the one we are concerned with here is a matter of the world being poised between various determinate states. In that sense, there are multiple determinate (precise) states between which the world is unsettled, such that it fails to specify which one obtains. For instance, Cameron (2015: 196) argues that the openness of the future consists in such brute unsettledness. He claims that if we think that the future is open with respect to whether or not there will be a sea-battle tomorrow, we ought to think that this is a matter of the world being unsettled as to which relevant state obtains.

From a semantic perspective, it is determinately the case that either a future contingent or its negation is true, but it is indeterminate *which* (where this indeterminacy is a brute unsettledness between ways the world could be), so that future contingents are said to be metaphysically indeterminate in truth-value (cf. Barnes & Cameron, 2009: 294). This is not to say that the worldly conditions required for the truth (or the falsity) of future contingents are absent, but rather that there is a 'lack of specificity' concerning *what* worldly conditions obtain. As an analogy, Barnes and Cameron take the borderline case of being bald: it might be indeterminate whether this person is bald or not, but it is determinate that *either* he is bald *or* he is not. Just as in the case of future contingents, it is determinate that the statement ⌜This person is bald⌝ is *either* true *or* false (those are the only two options), but it is indeterminate which of the two options is in fact the case (cf. Barnes & Cameron, 2009: 294).

The immediate benefit of such a characterization of the asymmetry is that it allows one to accept Bivalence without restriction and, therefore, to resist the non-classical logic and semantics introduced in Sect. 2.2 (three-valued treatment of truth-functional connectives, supervaluationism). Indeed, assuming that the indeterminacy in truth-value of future contingents results from the world being unsettled in this respect (one way or another, there is a fact of the matter), the world always speaks to the truth or falsity of any claim about how things will be. This allows one, for example, to assess as correct (or incorrect) some of our current and past assertions of statements about how things would turn out (assuming the orthodox account of assertion, cf. Sect. 2.2).[48] Moreover, since this characterization provides a robust understanding of the asymmetry as a genuine feature of how the world is, it avoids the main pitfalls encountered by the theories addressed so far. For example, it provides an explanation as to *why* our knowledge of the future is not as vast as our

---

[48] For example, suppose on Monday I make the prediction that there will be a sea-battle tomorrow. My assertion lacks a determinate truth-value. But come Tuesday, when the sea is fortunately peaceful, I can look back and say that my prediction was not correct. If it had been correct, a sea-battle would now be raging over the sea. It is not the case, so it was not correct. So, since it is determinate that no sea-battle is now raging, I can say that, determinately, my prediction that there would be a sea-battle was not correct. But I cannot say that it was determinately incorrect: it was not, because the future was open with respect to whether things would turn out as predicted (cf. Barnes & Cameron, 2011: 4).

knowledge of the past: we cannot know more about the future than what determinately will be the case.

However, despite the non-negligible benefits this characterization offers, there are some reasons to complain. First, this characterization faces difficulties in interpretation that render it less than metaphysically illuminating. It is unclear, for example, how metaphysical indeterminacy might involve an indeterminate degree of obtaining: whether or not a state obtains seems to be an all-or-nothing situation – either a state obtains or it does not. In particular, if the future is genuinely open, then intuitively, *none* of the relevant states 'obtain' – the future has not yet happened, after all. To clarify further, the 'metaphysical indeterminacy' account, since it analyses the openness of the future in terms of the world being 'stuck' between various states, *presupposes* that these states exist, while it is clear that all of them cannot share the same ontological status. After all, even assuming that both tomorrow's sea-battle and its peaceful alternative exist, only one of these states will be actualized. There must therefore be a difference between these two states: one must 'less obtain' than the other. As Barnes and Cameron recognize, the indeterminate degree of obtaining to which their account is committed covers two possibilities: "[…] perhaps the state of an object *indeterminately instantiating* a familiar property, or perhaps the state of an object instantiating the *non-familiar* property of *being indeterminately F*" (2017: 123). In other words, either the indeterminate degree of obtaining concerns the instantiation of the property, or the property itself. However, both possibilities are mysterious. Furthermore, whereas it is clear what it means for human beings to be *undecided* between various possibilities (e.g., to go, or not to go to the cinema), it is unclear what it means for the world to be *undecided* about, for example, what will happen tomorrow (e.g., there will be, or there will not be a sea-battle). Admittedly, there have been courageous attempts to make sense of this claim, as reflecting, for example, that it may be 'indeterminate which world is actualized' (Barnes & Cameron, 2011), or that there may be multiple 'actual worlds' (Williams, 2008a). But even if such attempts are coherent, they occupy, as Jessica Wilson says, "[…] a metaphysically tenuous region of logical space" (2013: 364).

Moreover, there are cases of future contingents lacking determinate truth-value that are not cases of the future being open. According to Barnes (2010), the conditions of being bald are such that there can be people for whom it is unsettled whether they are bald or not. If so, the future contingent ⌜This person will be bald tomorrow⌝ might *now* be metaphysically indeterminate in truth-value. But this does not necessarily seem like a case of the future being open. After all, what is going to happen could be perfectly settled. Perhaps the person in question will undergo a chemotherapy session tomorrow, which has been scheduled for months and which will inevitably make him lose a lot of hair (to such an extent that it will be unsettled whether he is bald or not). There are not different ways the future might develop here; there is only one settled future, but it is unsettled whether it will include an additional bald head. The analysis of the openness of the future as metaphysical indeterminacy might thus appear insufficient, since it fails to delineate cases of open future indeterminacy from other cases (e.g., cases of ontic vagueness). Admittedly, this objection could be prevented by denying that openness and vagueness are two phenomena of the same kind. After all, perhaps vagueness is merely a semantic deficiency of language that can be treated with, for example, a supervaluationist account of truth and validity (cf. Fine, 1975b). In other words, the above objection does not rule out the possibility that Barnes and Cameron might be wrong about vagueness, but right about the openness of the future.

However, it must be acknowledged that Barnes and Cameron (2011: 3) outline a solution to the latter issue. They claim that there is a clear delineation between cases of open future indeterminacy and other cases: the former cases will be resolved as time passes, while the latter will never be resolved. For example, while it might *now* be open whether or not there will be a sea-battle tomorrow, once tomorrow comes the situation will be resolved: depending on how the future unfolds, it will be determinate that there is a sea-battle or that there is not. By contrast, if it is *now* indeterminate whether or not a person will be bald tomorrow (assuming that the future is perfectly settled), once tomorrow comes it will *still* be indeterminate whether or not this person is bald. Indeed, assuming that it is indeterminate whether a person who has, say, 200 hairs on his head is bald or not, the passage of time does not help clear up the matter: tomorrow and the following days, it will *still* be indeterminate whether having 200 hairs is to be bald or not.

Yet, this solution may not appear very attractive, since it includes exceptions: (i) there are cases of open future indeterminacy that will never be resolved; (ii) there are cases of other sorts of indeterminacy that will be resolved. As a first example, consider the prediction ⌜Jesus will return someday⌝ and suppose for the sake of argument that it is open whether he will.[49] Suppose further that this prediction is uttered in a world where time never ends, and where, at any given moment, Jesus has not yet shown up. If time does in fact unfold this way then, it seems that the indeterminacy in 'Jesus will return' will never be resolved: for at all times, there will still be a future in which he might still return. After all, we could wait for the return of Jesus forever! According to Wilson, this kind of possibility is not straightforwardly handled on Barnes and Cameron's account, since in supposing that openness consists in the matter of the world being unsettled as to which relevant state obtains, "[…] it is presupposed that the future gets settled, one way or another" (2013: 381).

As a second example, consider the prediction ⌜Schrödinger's cat will be alive⌝ and suppose that it is uttered 5 min before opening the box. This prediction is *now* indeterminate in truth-value (the world fails to settle a unique determinate of the cat's life status). But, 5 min hence, when opening the box, ⌜Schrödinger's cat is alive⌝ will be determinately true or determinately false. This case will thus be resolved as time passes. However, it might be argued (although it is controversial) that this is not a case of the future being open. The indeterminacy is indeed to be explained by the weirdness of quantum mechanics, not by the passage of time. After all, one can easily imagine an alternative case in which, 5 min hence, the box will remain closed, so that the indeterminacy will not be resolved as time passes. Since

[49] This example can be found in Barnes & Cameron (2011: 5).

the second case, which arguably has nothing to do with the openness of the future, is of the same sort as the first one, it seems that the first case has nothing to do with the openness of the future either. It therefore appears that Barnes and Cameron fail to provide an absolute criterion to separate cases of the future being open from other cases and, therefore, that their account of openness in terms of metaphysical indeterminacy is insufficient (cf. also Wilson, 2016: 110–111).

Finally, there is an important sense in which the future may be said to be open but that the 'metaphysical indeterminacy' account fails to capture: time could come to an end, with no ontological commitment to future things standing in the way (cf. Correia & Rosenkranz, 2018: 99). Taking seriously the 'doomsday scenario' as it is described, for instance, in the cosmological 'Big Crunch' scenario (cf. Misner et al., 1973: 771), or in the eschatologies of the three major monotheisms (Judaism, Christianism, Islam), it could be that time does not go on indefinitely and, possibly, that there will be a last moment of time. In other words, it seems possible that the future is open not simply in terms of *how* it will unfold, but also in terms of *whether* it will unfold. However, Barnes and Cameron's account, in taking openness to be unsettledness between determinate states, presupposes that there are determinate states, and so cannot accommodate the possibility of radical openness, where more determinate states are simply not available because time has ended. To say that time could come to an end is to say that there could be no more determinate states for the world to be in, so that if the future may be said to be open in this radical sense, then Barnes and Cameron's account – which presupposes that there are determinate states – cannot accommodate it. By assuming that, one way or another, the future gets fixed, Barnes and Cameron rule out the possibility of radical openness.[50]

Of course, it might be argued that this latter objection only concerns Barnes and Cameron's account, and not all attempts to characterize the openness of the future in terms of metaphysical indeterminacy. After all, it seems intuitively possible to claim that in the same way that it is metaphysically indeterminate whether or not there will be a sea-battle tomorrow, it is indeterminate whether or not time will unfold. It might therefore be argued that another account of the openness of the future in terms of metaphysical indeterminacy could be better equipped to respond to the above objections. In particular, it might seem that abandoning the conception of time as being linear in favor of a conception of time as forward-branching could

[50] Setting Barnes and Cameron's account aside, it is possible to characterize the openness of the future in terms of metaphysical indeterminacy without being committed to the existence of any future state. For example, one could hold that while there is no future ontology, there are brute facts about what will happen, and that it is metaphysically indeterminate which of these brute facts obtain. However, although this option can allow for the radical sense in which the future may be said to be open, it is unattractive, since it leads to a *dilemma*. Either the past exists or the past does not exist. If the past does not exist, then it has to be treated as equally open (the asymmetry collapses). If the past does exist, then this option treats the fixity of the past and the openness of the future as being sensitive to different kinds of features of the world in a way that is *ad hoc*. Indeed, if the fixity of the past is, in some way or other, to be explained by the past ontology, then, due to considerations of parity, the openness of the future should be explained by the future ontology in just the same way.

express the idea of the metaphysically indeterminate future in a more appropriate way. Therefore, before taking up a final position on the 'metaphysical indeterminacy' accounts of openness, it is worth checking whether another type of worldly indeterminacy could meet the challenges presented so far.

2.8 The Branching Future

The idea of a non-linear time has been developed in various ways, but all the most plausible ones are based on the same principle: the past is actual and unique, while the future comprises many possibilities.[51] Intuitively, the temporal order of the world has the shape of a tree: the past constitutes a single trunk and the future a multiplicity of branches (each bearing the relation *is later than* to the present time). This difference in the topological structure of the future with respect to the past is meant to spell out the intuition that the future is open while the past is fixed. Indeed, whereas the linear conception of time admits at most one possible future for each time *t*, non-linear time permits instances in which a time *t* has many possible alternative futures. In that sense, non-linear time puts possible alternative futures into the topological structure of time, so that they must be considered in reckoning the truth-values of tensed statements (with the costly consequence that future contingents are neither true nor false) (cf. Thomason, 1970: 265).

For instance, 'the sea-battle case' can be sketched with such a tree model. It would comprise a trunk (C_0) and two branches, one representing the possibility that there will be a sea-battle tomorrow (C_1), the other representing the possibility that there will not be a sea-battle tomorrow (C_2). It would then be said to be open at (C_0) whether there will be a sea-battle, since a sea-battle occurs at (C_1) but not at (C_2). In other words, assuming that (C_1) and (C_2) are two *genuine* possibilities (neither is privileged over the other), it appears to be metaphysically indeterminate which of these two possibilities will be actualized and, therefore, how the future will unfold. Moreover, forward-branching time is arguably able to make sense of the radical possibility that time could come to an end by representing it by an absence of further branches from a node. But this is highly controversial. As, for instance, Ross Cameron claims: "[…] the absence of further branches from a node does not represent the further possibility that nothing will happen beyond that node, it simply represents the absence of further open possibilities" (2015: 179). Indeed, considering Fig. 2.1, two branches diverge from (C_0), thus signifying that there are two ways the future might unfold from (C_0). The absence of a third branch signifies that there are *only* two ways the future might unfold from (C_0), *not* that it might not unfold *any way* from (C_0) because (C_0) could be the last moment of time (cf. also Pooley, 2013: 340).

[51] Cf. Kripke [1958] 2011, Prior, 1962, 1967, Burgess, 1979, 1980; Thomason, 1970, 1984, Øhrstrøm & Hasle, 1995.

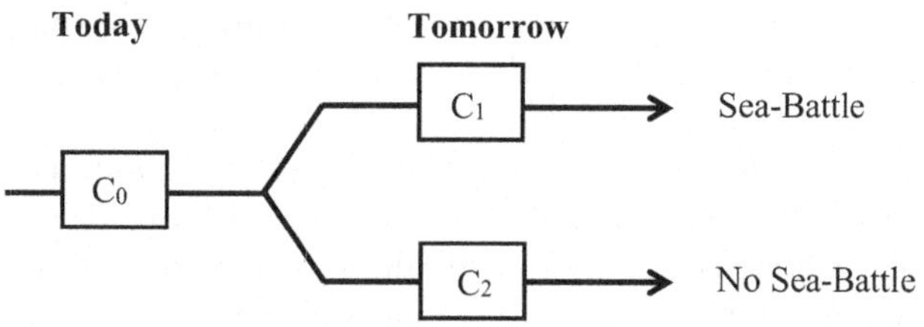

Fig. 2.1 Basic branching structure of time

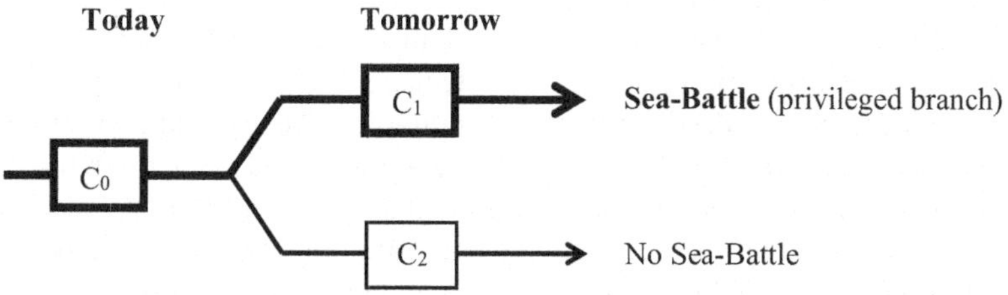

Fig. 2.2 Branching structure of time with a specification

Although branching time provides a forceful metaphor to picture the openness of the future, it might be denied that it captures a *genuine* notion of it. In particular, branching time appears like a "[…] a metaphysic on which it is perfectly settled how things will be, you just do not know whereabouts you will be within reality" (Cameron, 2015: 175). After all, at (C_0) it is perfectly settled that there is a sea-battle temporally beyond us. It only seems unsettled as to whether or not there will be a sea-battle because it is unsettled which of the two branches, (C_1) or (C_2), will be ours. But that is not a case of reality being unsettled; it is simply that, when we talk of the future, we commonly privilege a small portion of it, which corresponds to "[…] the branch we happen to find ourselves upon" (Cameron *id*). From a metaphysical point of view, there is thus nothing special about that branch; it is merely that it is *our* branch. It might thus be argued that genuine openness should not only arise as a result of our non-neutral perspective on time, but also requires reality to be truly unsettled (cf. Lewis, 1986: 207, Rosenkranz, 2013).

Of course, one might respond that *our* branch is metaphysically different from the others, e.g., by adding to the view a *specification* that determines *our* branch as the unique history of the world. This option may, in particular, come in two flavors: static and dynamic. According to the static option, it is not only true that either there will be, or there will not be a sea-battle tomorrow, but also that there is a present *specification* (commonly referred to as 'the thin red line') of what is actually going to happen (cf. Fig. 2.2). This specification breaks the symmetry between the two ways the future might develop, privileging the one that is actually going to be actualized. As a consequence, statements like ⌜There will be a sea-battle tomorrow⌝ are not exceptions to Bivalence: they *now* have a classical truth-value, albeit our limited minds keep us from knowing which (cf. Belnap & Green, 1994; Borghini &

Torrengo, 2013; Burgess, 1979; Thomason, 1970). Although the specification is to be taken as an objective feature of reality, it is not epistemically accessible: there is no way to know which possible future will be actualized, except by waiting. Arguably, this option is compatible with believing that the future is open, since it retains the crucial idea that even if there will *in fact* be a sea-battle tomorrow, *there need not be*: even though one of two possibilities (either there will be, or there will not be a sea-battle tomorrow) is already the one which is going to be actualized (we cannot know which one), it *could have been* the other.

However, many philosophers reject the idea that a privileged future possibility is compatible with the open future. For example, Belnap and Green (1994), MacFarlane (2003), and Barnes and Cameron (2011) argue that if one wants to hang on to *genuine* openness regarding the future, it is problematic to give one future possibility a special status. It might indeed seem that the intuition of the open future is more radical than what is supposed in the static option. As Barnes and Cameron write: "[i]t's not that there is a way our future actually is but that there are different ways it *could have* been [...]; it's that there actually are now multiple possible ways our future could turn out to be" (2011: 2). After all, from a God's eye point of view, looking down on the tree and seeing that only one possibility remains in play (i.e. the branch singled out as that which is going to happen), the other possibilities are not *genuine* possibilities. They merely represent possibilities in an epistemic sense: we do not know which one will be actualized.

By contrast, the dynamic option seems to capture a *genuine* notion of openness. The idea is to claim that the tree changes as time progresses, e.g., by adding to the view a process of 'branch attrition' – the branches which are ruled out as being part of history cease to exist (cf. Fig. 2.3). This idea has, in particular, been defended by Storrs McCall (1994). According to his view, the passage of time is represented by the vanishing or 'falling off' of branches; the one branch remaining being the 'actual' one that becomes part of the trunk. The probability of any future event (e.g., tomorrow's sea-battle) is specified by the proportion of future branches on which that event occurs. However, although this view pictures a genuine notion of openness (branch attrition depicts the past-future asymmetry in an objective, observer-independent way), it is unsatisfying. First, 'branch attrition' is not a physical mechanism and, as far as I know, there is no physical mechanism that can account for it (but see quantum decoherence). It therefore seems that McCall's model fails to explain why, for example, we cannot climb down the tree trunk and return to the

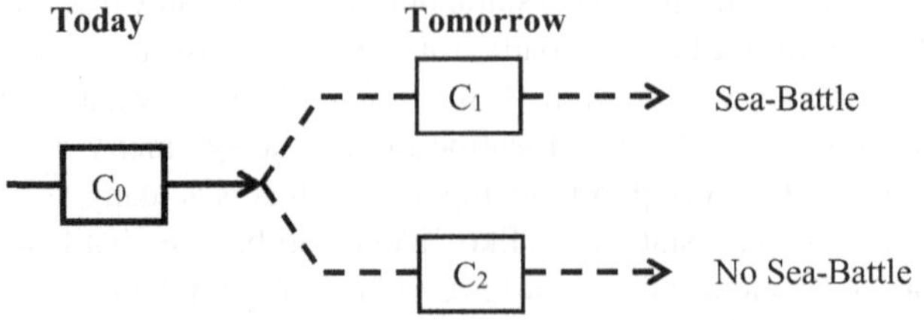

Fig. 2.3 McCall's dynamic branching structure of time

past. Second, contrary to what is entailed by McCall's view (that there is no branch that *used* to diverge from earlier times, since these branches ceased to exist when those times became past), we do not think of the present as the first time from which there is branching. We rather think that in the past there were multiple ways history could have gone but did not.

But if reality was indeed open prior to now, despite the fact it is not branching before now, what allows us to exclude the possibility that reality might be closed in the future, despite the fact that it is branching from now on? It seems that as soon as McCall admits that there are resources (e.g., brute tensed facts about how the tree used to be) to say how the branching reality *was*, he has to allow the possibility that there are resources about which branch our history *will be* and, therefore, that the future is perhaps not open after all (despite the multiplicity of branches).[52] Thus, although branching trees might seem convenient as long as we are concerned with the semantics of future contingents (cf. Belnap, 1992; Belnap et al., 2001; Prior, 1967; Thomason, 1970, 1984), they fail to provide a perspicuous picture of the metaphysics underlying the asymmetry in openness between the future and the past.

Another reason *why* dynamic branching-tree models do not seem appropriate to account for the asymmetry between the open future and the fixed past has to do with the logical possibility of time-travel. Specifically, as soon as time-travel enters the picture, it seems that the openness of the future cannot be characterized by means of alternative future branches. To illustrate this point, consider the following scenario. Suppose that t_1 is the objective present and that only two future branches stem from t_1: (C_1) and (C_2). Max is born at t_2 on (C_1), whereas Mary is born at t_2 on (C_2). Assuming that time-travel is possible, we can imagine that both Max and Mary have a time-machine, and hence that both of them can travel to the past. Now, suppose that both Max and Mary travel back in time to t_1 from their respective locations. According to the dynamic branching-tree model, time moves forward and only one of the two branches, (C_1) or (C_2), becomes actual. Let us suppose that (C_1) is the branch that becomes actual, and thus that (C_2) drops off. The branch where Mary is born and activates her time-machine does no longer exist. What happens at t_1 then?

There are only two possibilities which both lead to difficulties: either (i) Mary exists at t_1 when t_1 is present, but does not exist at t_1 when t_2 is present, or (ii) Mary tenselessly exists at t_1. The first possibility contradicts the principle according to which it is not possible to change the past (t_1 change from containing Mary (when t_1 is present) to *not* containing her (when t_2 is present)); the second possibility contradicts (at least when t_2 is present) the Lewisian principle according to which P is a time-traveler only if all P's temporal stages are united by some causal relation (since Mary's t_1 person-stage is not causally related to any person-stages in the future) (cf.

[52] Of course, McCall can simply deny that there are such resources, but then the reasons for the future's being open is not simply that there are multiple branches; it is that, together with there being no resources as to which branch will be actualized. Openness is no longer being characterized by variation across the extant branches, and so the branching view fails to accommodate the asymmetry.

Lewis, 1976: 148).[53] Since we might not want to give up any of these two plausible principles, we might conclude that dynamic branching-tree models of time are incompatible with time-travel (cf. Andreoletti, 2020; Miller, 2005; Norton, 2018). However, as we will see in the fourth chapter (Sect. 4.7), a natural way to resist this kind of objection is to accept, for instance, a revised conception of perdurantism.

This new failure in properly characterizing the asymmetry between the open future and the fixed past leads us to think that this requires another type of worldly unsettledness that is not to be expressed in terms of the world being indeterminate whether something is the case (neither by postulating multiple states that the world is unsettled between, nor by postulating multiple forward branches), but rather in terms of there being *no fact of the matter* whether something is the case. Before introducing this last characterization, I would like to address an objection that might be raised against both the 'metaphysical indeterminacy' and the 'no fact of the matter' account. This objection relies on the thought that worldly unsettledness may not be a *fundamental* phenomenon, but merely a *derivative* one (cf. Bacon, 2019; Eva, 2018; Sattig, 2014). At first sight, this objection seems absurd: if there is no unsettledness in fundamentals, there is nowhere for unsettledness to come from. As Elizabeth Barnes puts it: "[…] if you combine a bunch of determinate things, you won't (no matter how you combine them) be able to get any (genuine) indeterminacy" (2014: 341). However, the thought that 'if reality is really unsettled, then it must be in virtue of unsettledness in fundamentals' rests on two principles that can be disputed: (i) fundamental facts ground the derivative, (ii) grounding is determinacy-preserving. In particular, it might be argued that grounding (or the link between the fundamental and derivative, whatever it is) can itself be unsettled. For example, one might think that the fundamental facts do indeed ground the derivative ones, but since it can be unsettled *what* derivative facts the fundamental facts ground, the facts about what grounds what are themselves unsettled. An immediate reply to this objection, however, is that the facts about what grounds what belong to the fundamentals. After all, if these facts were not fundamental, they would in turn be grounded in some other facts in a way that seems to involve a vicious regress.[54]

---

[53] This principle demands the proper kind of causal connectedness among the stages of a time-traveler in order to rule out cases of counterfeit time-travel. As David Lewis observes, if an individual is *randomly* created by a demon out of thin air at a time t and it happens to be a duplicate of a stage of a person P annihilated at a later time t_1, this should not count as a case of time-travel: there is no causal continuity among the stage disappearing at t_1 and the stage appearing out of nowhere at the earlier time t (cf. Lewis, 1976: 148). It is worth noting that the second possibility also contradicts the principle according to which it is not possible to change the past, since Mary changes from being a genuine time-traveler (when t_1 is present), causally connected to her earlier temporal stages, to *not* being a genuine time-traveler (when t_2 is present), having lost her earlier temporal stages.

[54] This claim has been challenged by Sider, 2011, Bennett, 2011, and Derosset, 2013. For instance, according to Ted Sider's principle of *purity*, fundamental truths involve only fundamental notions, such that facts about what grounds what (which explicitly refers to the non-fundamental level) cannot count as fundamental.

So, even if principle (ii) is false (which is doubtful), then – given that the facts about what grounds what are fundamental – the fundamental level is the source of unsettledness.

2.9 No Future!

When Johnny Rotten of the Sex Pistols shouts "No future!" (in the closing refrain of the controversial song "God Save the Queen"), what he probably intended to say is that the future will certainly be unfair for the English working class. Nonetheless, this slogan (which has since become emblematic of the punk rock movement) might be interpreted in a more literal and, therefore, radical sense: the future is simply nothing at all. This is precisely what C. D. Broad defended, long before the advent of punk rock music. According to his doctrine, "[n]othing has happened to the present by becoming past except that fresh slices of existence have been added to the total history of the world. The past is thus as real as the present. On the other hand, the essence of a present event is, not that it precedes future events, but that there is quite literally *nothing* to which it has the relation of precedence. The sum total of existence is always increasing, and it is this which gives the time-series a sense as well as an order" (1923: 66–67). In Broad's view, the present is thus a kind of 'ontological gateway' through which events have to pass on their way to become real and always remain so.

Although going beyond Broad's view is surprisingly difficult (this will occupy us throughout the next chapter), his intuitions may provide decisive insights into the temporal asymmetry we are concerned with. Indeed, perhaps the asymmetry between the open future and the fixed past is to be characterized by an ontological difference: the past and the present exist, while the future does not exist. In that sense, the openness of the future (as a kind of worldly unsettledness) should perhaps not be regarded as a matter of the world being poised between various determinate states (as Barnes & Cameron suggest), but rather as there being no state that the world is in with respect to what will happen. For example, if we think the future is open with respect to whether or not there will be a sea-battle tomorrow, we ought to think that reality simply lacks the relevant ontology: there is no fact of the matter whether a sea-battle will take place tomorrow. Mauro Dorato comes to similar conclusions when he comments on this passage from Broad's book. He writes: "[t]he main thought here seems to be that by leaving the future wholly *empty*, we make sense *both* of the fact that our actions can give a (cosmically negligible) contribution to bring it about, *and* of our closely related intuition, hard to explicate in a clear way, that at any instant of time, there is a part of the history of the universe that is 'fixed' and 'definite', and a part that isn't" (2008: 56). Besides metaphysical indeterminacy, there is thus another way in which reality may fail to settle some future facts: there may be *no fact of the matter* whether these facts obtain.[55]

These two types of worldly unsettledness seem to be very different *by nature*: the first type is a case of *overdetermination* (there are too many states, such that the world is unsettled as to which one obtains), while the second type is a case of *underdetermination* (there is no state, such that the world is unsettled as to whether something obtains). As an analogy, consider two teenagers – one rich, the other poor – getting ready for a party. In front of their closets, they might both be 'unsettled' as to what to wear for the evening: the rich teenager because she is spoiled for choice (e.g., she has too many dresses suitable for the occasion), and the poor teenager because he has *literally* nothing to wear. In such a case, although the two teenagers are equally unsettled as to what to wear for the party, it clearly appears that this reflects two different phenomena.

The second kind of unsettledness, by the way, is commonly invoked in many different contexts. In the philosophy of quantum physics, for example, the superposition of states as a source of metaphysical indeterminacy can successfully be accommodated by a so-called determinable-based account, which treats certain sources via a gappy implementation. For example, as has already been mentioned, Calosi and Wilson (2019) argue that the case of Schrödinger's cat suggests a metaphysical indeterminacy that must be seen as involving that there is *no (determinate) fact of the matter* whether the cat is or is not alive. Specifically, there is no fact such that the property of *having a certain life status* of Schrödinger's cat has a unique determinate (*being alive* or *being dead*). Thus, as in the case of the open future, the determinable-based account of quantum indeterminacy does not involve that it is indeterminate which of various determinate facts obtain (*metaphysical indeterminacy*), but rather that it is determinate that no determinate fact obtains (*no fact of the matter*).

Of course, this account is *insufficient* to accommodate the open future, since the future is intuitively fixed in a fully deterministic world (where the state of the world at a time nomologically necessitates the state of the world at any later time). As has previously been argued, if it is *necessary*, given the facts about how things are up to a time *t* and what laws obtain, that the world will be a certain way at any later time, then it seems *settled* that it will be that way, which is just to say that the future is *fully settled* at *t* and, therefore, that the future is fixed. However, assuming that *all* that will happen is not made inevitable by how the world currently is (together with the laws of nature) – which sounds quite plausible in regard of important results of contemporary physics (especially quantum mechanics) – the 'no fact of the matter' account provides a powerful characterization of the openness of the future, understood as a kind of worldly unsettledness.

[55] Of course, someone who denies the existence of the future is not forced to accept the 'no fact of the matter' account of openness. He is not even forced to say that the future is open in any sense whatsoever: he could hold, for example, that while there is no future ontology, there are brute facts about what will happen (cf. Cameron, 2015: 194–195).

From a semantic point of view, it is not that there is a 'lack of specificity' concerning *which* worldly conditions for future contingents obtain, but rather that worldly conditions are absent: there is no fact that speaks to the truth or falsity of ⌜There will be a sea-battle tomorrow⌝. This kind of unsettledness might thus seem to be incompatible with Bivalence. This is, at least, what Broad himself contends when he claims that future contingents are without exception neither true nor false (cf. 1923: 70–73). After all, if there is no fact of the matter as to whether Fφ (where 'Fφ' stands for 'It will be the case that φ'), i.e. if nothing worldly answers to whether or not Fφ, it might seem that Fφ should be neither true nor false. In particular, Fφ should not be true, because the worldly conditions required for its truth are absent, but nor should it be false, because the worldly conditions required for its falsity are also absent. It might thus be concluded that if there are some statements concerning which there is no fact of the matter, Bivalence must be rejected (cf. Cameron, 2015: 181).

This conclusion is clearly problematic. As has already been said, most philosophers believe that "[...] classical semantics and logic are vastly superior to the alternatives in simplicity, power, past success, and integration with theories in other domains" (Williamson, 1994: 186). However, it is not clear that this has to be accepted. According to Correia and Rosenkranz (2018), this conclusion is driven by too strong a conception of the 'grounding requirement on tensed truths' (i.e. the requirement according to which tensed truths do not 'float free', but are grounded in reality).[56] They argue that the present truth of a given statement does not require it to be grounded in how things located in the present are, but that it might well be grounded, at some future time, by things whose future existence and future ways of being are such as the statement claims. For example, the statement ⌜There will be a sea-battle tomorrow⌝ might well be true *now*, provided that, 1 day hence, there will be a sea-battle the existence of which will explain *why*, 1 day before, the statement was true.

This theoretical option allows Correia and Rosenkranz to affirm that future contingents are not exceptions to Bivalence – they *now* have a classical truth-value – while keeping the future open. Indeed, assuming that there will be things whose existence is not rendered inevitable by how things located in the present or past of now are or were, the present truth of a statement that will be grounded by how matters are going to stand does not undermine its status as a future contingent. In other words, the bivalence of statements about the future that will be grounded by what there will be and how it will be does not threaten the open future as long as "[…] nothing there is or was, in conjunction with how it is or was, makes it inevitable that, in the future, there will be such grounds" (2018: 110). Thus, through a weaker (and more plausible) conception of the grounding requirement on tensed truths – "[t]he truth of a given tensed statement at most requires that it *sometimes* be grounded in

[56] The grounding requirement on tensed truths, mentioned by Correia and Rosenkranz, is a stronger principle than the one accepted so far: for p to be true, the worldly conditions for p's truth would have to obtain; and for p to be false, the worldly conditions for p's falsity would have to obtain. However, this plays no essential role in the discussion.

what then is something and a certain way" (2018: 108) – Correia and Rosenkranz's option allows one to reconcile the bivalence of future contingents with the 'no fact of the matter' account of openness: assuming physical indeterminism, the future may be said to be open in the sense that there is no fact of the matter regarding the grounds of some present truths about the future, without future contingents lacking a classical truth-value.

Conversely, this weaker conception of the 'grounding requirement on tensed truths' allows one to account for the truth of statements about the past, without committing to the existence of the past. Indeed, since the truth-value of a tensed statement does not need to be grounded in how things located in the present are (e.g., a future contingent might well be said to be true *now* in virtue of how, at some future time, things will be), a past-tensed statement might well be said to be true *now* in virtue of how, at some past time, things have been. For example, the truth of the statement ⌜Napoleon lost at Waterloo⌝ merely requires that things have been that way, i.e. that Napoleon has actually been defeated by the Seventh Coalition. The weaker conception of the 'grounding requirement on tensed truths' therefore renders the existence of the past *superfluous* to ground present truths about it: that the Battle of Waterloo, located in the past, still exists, is *superfluous* to ground the present truth of ⌜Napoleon lost at Waterloo⌝. As a consequence, the metaphysicians who deny the existence of the past (e.g., the presentists) are in no worse position than those who do not (e.g., the growing blockers) to provide present grounds for truths about the past. To put it another way, since the weaker conception of the 'grounding requirement on tensed truths' is available to both presentists and growing blockers, they are equally well positioned to account for truths about the past. Resorting to the existence of the past to obtain grounds (truthmakers) for present truths about the past (as, for instance, some growing blockers do) is therefore useless. Thus, while one might have various independent reasons to defend the existence of the past (e.g., to ensure its fixity), the semantics of tensed statements should not count as one of them.

However, it must be acknowledged that this weaker conception of the grounding requirement on tensed truths is often regarded with suspicion. In particular, a question that constantly recurs is: 'How can a statement that has not yet been made true *be true*?' What lies behind this question is presumably the belief that at least some statements have no truth-value until they are made true (or false) by actions and other events. It is indeed often assumed that many of our statements about the future express propositions that are neither true nor false when they are made, but become true or false when events make them so in the future (cf. Broad, 1923: 70–73). For example, it might seem that the statement ⌜Switzerland is the next champion of the football world cup⌝ is not true (or false) until the events of the next world cup (e.g., until the delivery of the cup). In the next few lines, I will argue that this belief rests on a confusion between two different properties having to do with truth – *being true*, and *being made true* – and should therefore be abandoned.[57]

First, it is worth noting that the converse of the belief in question does not hold: it is not true that if a statement is true, then some events either have made, are making, or will make it true. There are important statements whose truth-value is established not by being made true by events, but in some other way. For example, consider Pythagoras's theorem: no event has ever made true the statement ⌜The square of the hypotenuse of a right triangle is equal to the sum of the squares of the other two sides⌝, and none ever will. It is not like the statement ⌜Biden won the 2020 election⌝. There is no sequence of events, ending at a certain time, the occurrence of which makes Pythagoras's theorem true. Perhaps it would be correct to say that, although Pythagoras's theorem is *independent* of events, events do *conform* to it. But, if so, events do not *make* this theorem true; they merely *reflect* its truth (cf. Perry, 2004: 235–236).

Second, suppose that the laws of nature are not made true by events, and that these laws, together with statements made true by events that have already happened, entail that Sam will not go to the cinema at some future time *t*.[58] It seems that, although the statement ⌜Sam will not go to the cinema at *t*⌝ has not yet been *made* true, its truth-value has been *settled*. Of course, the statement will not be *made* true until the events that are determined by the laws of nature and the past have actually occurred, i.e. until Sam goes to bed instead of going to the cinema at time *t*. But, as Perry puts it: "[…] these events were already entailed by a combination of propositions some of which were *already* made true and the rest of which aren't made true by events at all" (2004: 236). Therefore, against what skepticism about the weaker conception of the grounding requirement suggests, it clearly seems that the truth-value of the statement ⌜Sam will not watch a movie at *t*⌝ was settled *before* Sam refrained from going to the cinema.

Presumably, what is wrong with the belief that statements are not true until they are made true is that it rests on a confusion between two properties having to do with truth of statements: *being true*, and *being made true*. The former is a timeless property of statements (i.e. a property that is not relative to times), while the latter is a property that occurs at times, or through intervals. Although these two properties must be distinguished, this does not mean that they are not somehow related. Intuitively, if a statement is ever *made* true, it *is* true (where '*is* true' is to be understood timelessly). However, the fact that a statement has not yet been made true by events, does not imply that it *is* not true. John Perry proposes the following analogy: "[t]he fact that [Kamala Harris] has not yet been chosen as our next President does not imply that [she] is not our next president" (2004: 235). If Harris ends up being

[57] I will leave aside the argument that the belief in question does not mesh easily with the two-valued logic that most of us find convenient to work with, since this argument has already been developed (cf. Sect. 2.2).

[58] Under this hypothesis, laws of nature are not merely true generalizations of events (*pace* Hume); they are true generalizations that derive from the nature of things, and so describe constraints that form the structure of the world.

nominated and elected in 2024, then she is the next president of the United States. If I call her 'the next president of the US' now, I am correct if the future goes one way, incorrect if it goes the other. After all, if Harris must not be the next President of the US (on the grounds that it has not been decided yet) then, by parity of reasoning, no one is the next president of the US, which will be a "constitutional crisis"! To avoid the crisis, Perry argues, one must acknowledge that being the next president of the US is a property Harris has if, at some point between now (June 2022) and November 2024, she is elected as the next president. Perry concludes: "[t]he fact that lots of propositions be true that have not yet been made true is sort of like that" (2004: 235). Assuming that Perry is right, it seems that a statement, such as ⌜There will be a sea-battle tomorrow⌝, can be true *now*, partly because of contingent events that have yet to occur.

Thus, when combined with Correia and Rosenkranz's relaxed conception of the grounding requirement on tensed truths, the 'no fact of the matter' account seems to provide a powerful answer to the question of the open future, at least assuming that physical determinism is false. First, it enables us to preserve a classical logic, and in particular a bivalent semantics, even for future contingents. Second, it enables us to preserve the bivalence of future contingents without the future being bound to be a certain way. In that sense, although the statement ⌜There will be a sea-battle tomorrow⌝ is *determinate* in truth-value, because 1 day hence, there will be things whose existence will explain *why* this statement is true *now* (or false *now*), it is not *determined* to be true or to be false, because (i) there is no fact of the matter as to whether a sea-battle will take place tomorrow, and (ii) the current history of the world in conjunction with the laws of nature does not necessitate that a sea-battle will take place tomorrow. One therefore ends up with a substantial account of the open future, conceived as a kind of worldly unsettledness, which allows for unrestricted Bivalence, and hence resists the non-classical logic and semantic introduced in Sect. 2.2.

However, so far it has only been shown that the 'no fact of the matter' account of openness does as well as the 'metaphysical indeterminacy' account. No reason for privileging one account over the other has yet been provided. For instance, it is not clear that the latter account is more successful in delineating cases of open future indeterminacy from cases of other sorts of indeterminacy (e.g., cases of quantum metaphysical indeterminacy). After all, a statement like ⌜Schrödinger's cat will be alive⌝ may also be said to be determinate in truth-value (since, when opening the box, there will be a cat whose life status will explain *why* this statement is true *now* (or false *now*)), while it is not *determined* to be true or to be false (since (i) there is no fact of the matter as to whether Schrödinger's cat will be alive (or dead), and (ii) the current history of the world in conjunction with the laws of nature does not necessitate that Schrödinger's cat will be alive (or dead)).[59] This invites us to answer

---

[59] However, there might still be a way to delineate open future indeterminacy (conceived as there being no fact of the matter) from other sorts of indeterminacy (especially quantum metaphysical indeterminacy): whereas open future indeterminacy involves that there is no relevant ontology at all (there is simply no future), quantum indeterminacy merely involves that there is no *determinate* ontology (there are states of affairs whose constitutive entities have determinable properties but no unique determinate of these properties). The openness of the future might thus be singled out as being the only sort of indeterminacy that presupposes a *real lack* in the ontology.

the following question: is there any reason for preferring the 'no fact of the matter' account to the 'metaphysical indeterminacy' account?. The answer is 'yes'; there are at least two reasons that put the former in a better position than the latter: (i) the 'no fact of the matter' account is more metaphysically illuminating; (ii) the 'no fact of the matter' account is more powerful in capturing a radical sense of openness. Let me develop these two reasons.

First, whereas the 'metaphysical indeterminacy' account accommodates the openness of the future by introducing an obscure indeterminate degree of obtaining (openness involves indeterminacy about which determinate state obtains, cf. Sect. 2.7), the 'no fact of the matter' account accommodates it by simply denying that any future state obtains. Not only does this second option appear to be decidedly clearer (how should indeterminacy in obtaining be interpreted?), it also seems to be closer to the commonsense view. Indeed, the commonsense view according to which the future is open seems to be naturally interpreted as expressing relevant states as *determinately failing* to obtain, not as being such that it is indeterminate which relevant determinate states obtain. In particular, if the future is genuinely open about tomorrow's sea-battle, then intuitively, *neither* tomorrow's sea-battle *nor* its peaceful alternative obtains – the future has not yet happened, after all. It therefore seems that the 'no fact of the matter' account is superior to its rival not only in intelligibility, but also in how it tallies with the way we commonly think of the future.

Second, as we have seen, there is a sense in which the future may be said to be open that the 'metaphysical indeterminacy' account fails to capture (since it presupposes that the future gets settled, one way or another): time could come to an end, with no ontological commitment to future things standing in the way (cf. Correia & Rosenkranz, 2018: 99). Yet, there are good reasons to think that the 'no fact of the matter' account is better positioned for capturing this radical sense of openness. Indeed, given that this account presupposes a gap in ontology (i.e. there is no future), it seems that, assuming physical indeterminism, it can allow for such a doomsday scenario. After all, the possibility that time could come to an end is no more than a possible interpretation of what a strong form of physical indeterminism can lead to. Considering the world history up to t (especially all the entities actually existing at t, or at any time earlier than t), it might be that any time later than t never exists. Supposing that the future may be said to be open in this radical sense, it clearly appears that only an account of openness that does not presuppose that there are times later than t can allow for it. There therefore is at least one sense in which the future may be said to be open which coheres with the 'no fact of the matter' account, but not with the 'metaphysical indeterminacy' account. It can therefore be concluded that the 'no fact of the matter' account of openness is more powerful than its main rivals, and hence that, assuming physical indeterminism, the intuitive asymmetry between the open future and the fixed past should be characterized in ontological terms: there being facts of the matter about what happened, but not about what will happen.

2.10 Conclusion

In this chapter, I showed that there are various ways in which the asymmetry reflected by our intuition of an open future and a fixed past can be characterized. I argued that the substantial characterizations of the asymmetry (the asymmetry reflects how the world truly is) are more promising than the perspectival characterizations (the asymmetry merely reflects how humans interact with world), as the latter fail to provide a fundamental explanation (i.e. not causal or thermodynamic) of the intuitive asymmetry. Considering physical indeterminism, I argued that, though necessary, this doctrine is insufficient for characterizing the openness of the future and must, therefore, be combined with a metaphysical account of unsettledness. In this last respect, I presented two very different accounts of how reality may fail to fully settle what will happen: (i) although there are facts about what will happen, it is indeterminate which of these facts will obtain (*metaphysical indeterminacy*), (ii) there is simply no fact of the matter about what will happen (*no fact of the matter*). Following Correia and Rosenkranz (2018), I argued that the latter account can be reconciled with the bivalence of future contingents (without settling how the future will be), provided there is an appropriately relaxed conception of the grounding requirement on tensed truths. Finally, I showed that the 'no fact of the matter' account of openness should be preferred, since (i) it is more metaphysically illuminating and (ii) it coheres with a radical sense in which the future may be said to be open that is unavailable to the 'metaphysical indeterminacy' account. The various accounts appear in Fig. 2.4.

	Perspectival characterizations		**Substantial characterizations**				
	Semantic	Epistemic /Anthropo -centric	Physical	Modal	Metaphysical	Non-linear	Ontological (+ physical indeterminism)
Bivalence	No	Yes	Yes	Yes	Yes	No	Yes*
Explanatory power	Weak	Weak	Weak	Weak	Good	Good	Good
Intelligibility	Good	Good	Weak	Weak	Weak	Good	Good
Radical openness	No	No	Yes**	Yes	No	No	Yes

*with a relaxed conception of the grounding requirement on tensed truths
**with a non-futurist ontology

Fig. 2.4 The main features of the perspectival and substantial approaches to the asymmetry

Chapter 3
A Model for the Asymmetry

Abstract As has been argued, there are good reasons to think that, assuming physical indeterminism, the asymmetry between the 'open future' and the 'fixed past' is to be characterized as a kind of worldly unsettledness: there being facts of the matter about what happened, but not about what will happen. However, the main models of the temporal structure of the world – eternalism and presentism – do not reflect any ontological asymmetry between the future and the past. According to these models, either both the future and the past exist, or neither the future nor the past exists. So, in this chapter, I argue that we should opt for an alternative model of the temporal structure of the world – the growing block theory (GBT) – that seems better designed to accommodate the asymmetry in openness between the future and the past.

3.1 Introduction

In the previous chapter, it was argued that there are good reasons to think that, assuming physical indeterminism, the asymmetry between the open future and the fixed past is to be characterized as a kind of worldly unsettledness: there being facts of the matter about what happened, but not about what will happen. This characterization seems indeed to be required to fully account for the various ways in which our intuition that the future is open and the past fixed may be expressed. In particular, the radical sense of openness in which time could come to an end (with no ontological commitment to future things standing in the way) can only be captured by an account that *presupposes* a real gap in ontology (there is no future). However, the main models of the temporal structure of the world do not reflect any asymmetry between the future and the past. According to eternalism and presentism, the future and the past are ontologically on a par. Eternalists hold that both the future and the past exist, while presentists hold that neither the future nor the past exists. In other words, the two main competing models of the temporal structure of the world do not

This chapter (especially Sects. 3.6, 3.7, 3.8, 3.9 and 3.10) is based on Grandjean (2021b, c) and Grandjean (2022) but contains new material.

V. Grandjean, *The Asymmetric Nature of Time*, Synthese Library 468,
https://doi.org/10.1007/978-3-031-09763-8_3

ontologically distinguish the future from the past (either both of them exist or none of them exists). Therefore, neither eternalism nor presentism seems able to accommodate the asymmetry reflected by our basic intuition regarding the nature of time.

This conclusion leads us to think that we should opt for another model of the temporal structure of the world that provides an ontological ground for the asymmetry in openness between the future and the past. In that respect, the growing block theory (GBT), most famously put forward by C. D. Broad (1923), seems to be a natural candidate. This theory is indeed committed to the existence of the past (and the present), but not to the existence of the future. It depicts the block universe as increasing, with new moments of times coming into existence to join the moments that already exist. However, since GBT is commonly defined as a hybrid between eternalism and presentism (the growing blocker agrees with the eternalist that the past exists and agrees with the presentist that the future does not exist), it is often criticized for accumulating the flaws that are identified in the two traditional models. For example, just like eternalism (at least in its tense-realistic version), GBT seems to face with the so-called 'epistemic objection' (according to which there is no way of knowing that our time is the objective present) and, just like presentism, GBT seems to conflict with the theories of relativity (e.g., by requiring an absolute notion of objective simultaneity).

Therefore, in order to provide a defense of GBT, I argue that the traditional way of defining the A-theories of time – in terms of whether the future and the past exist – should be abandoned. To make that case, I argue that whereas nothing in actual intuition answers to the question 'Do the future and the past exist?', we clearly intuit that the future *differs* from the past (e.g., causes occur prior to their effects, the arrow of time points from past to future, the future is open while the past is fixed, etc.). My proposal in this chapter is therefore to make a fresh start in the debate by distinguishing two groups of A-theories of time: the *asymmetric* theories that have *intrinsic* resources to explain *why* (at least some) differences between the future and the past are intuited, and the *symmetric* theories that need *extrinsic* resources (e.g., entropy, irreversibility). I then distinguish the various forms *asymmetric* and *symmetric* theories may adopt by introducing two dynamic principles: *Temporal Becoming* and *Annihilation*. One immediate consequence of this proposal is that, contrary to what traditional definitions suggest, it is not *essential* to GBT to be an ontological hybrid between the polar opposites of eternalism and presentism. Rather, what is *essential* to this theory (at least in its full form) is to be *asymmetric*, while accepting *Temporal Becoming* and rejecting *Annihilation*. In this revisited picture, GBT does not represent an intermediate view, but a real alternative: assuming that some intuitive past-future time differences reflect the deep structure of reality, GBT is better positioned than its rivals to accommodate them.

The chapter is structured as follows. In the second section, I present the traditional way of introducing the various A-theories of time; I start from the McTaggartian distinction between the A- and the B-series, two ways of ordering positions in time, and explain *why* they both lead McTaggart to think that time is unreal. Then, using Kit Fine's reconstruction of the argument, I show that various ways of rejecting the principle of *Neutrality* allow one to derive various A-theories of time. In the third section, I express my dissatisfaction regarding the traditional way of defining the A-theories of time, since it neglects the role our intuitions are meant to play in the debate. In the fourth section, I make a fresh start in the debate by introducing two questions the specific answers to which will allow us (i) to get a comprehensive categorization of the A-theories of time and (ii) to reveal the metaphysical singularity and potential of GBT. In the fifth section, I focus on GBT and argue that, combined with physical indeterminism, it provides a solid basis to accommodate the intuitive asymmetry between the open future and the fixed past. In the sixth section, I present what is generally taken to be one of the most prominent objections against GBT, commonly referred to as the 'epistemic objection'. In the seventh section, I review various unsatisfactory attempts to address the epistemic objection, in particular Merricks (2006), Forrest (2004), and Correia and Rosenkranz (2018). In the eighth section, I show *why* the epistemic objection naturally leads to an anti-essentialist picture of kinds. In the ninth section, I show *how* an intuitive appeal to the continued existence of bare particulars may help solve the epistemic objection and, more generally, may account for *how* things located in the past exist. Finally, in the tenth section, I justify the claim that there are such bare particulars.

3.2 The McTaggartian Picture

In his best-known paper published in 1908, J. M. E. McTaggart argues that time is unreal, because our descriptions of time are either *insufficient*, *contradictory* or generate a *vicious circle* or an *infinite regress*. McTaggart begins his argument by distinguishing two ways of ordering positions in time. First, positions can be ordered according to their possession of properties like *being two days past*, *being present*, *being one day future*, etc. – these properties are often referred to now as 'A-properties'. McTaggart calls the series of positions ordered by these properties the 'A-series'. Second, positions in time can be ordered by two-place relations like *two days earlier than*, *simultaneous with*, *one day later than*, etc. – these relations are often referred to now as 'B-relations'. McTaggart calls the series of positions ordered by these relations the 'B-series'.[1] This distinction between the A-series and the B-series has served as a natural starting point for most of subsequent work on the metaphysics of time. In particular, it has offered a framework within which the main models of the temporal structure of the world have since been developed.

Nowadays, philosophers of time are said to hold an 'A-theory of time' or a 'B-theory of time', depending upon their attitudes to the A-properties and B-relations. These labels, first coined by Richard M. Gale (1966), can be understood as follows: the 'A-theorists' (or 'tense realists')[2] claim that there is an objective

---

[1] While both of these series are essential for the reality of time, the A-series is the more fundamental of the two, since the B-series can be derived from it alone (cf. McTaggart, 1908: 463).

[2] Cf. Bigelow, 1996; Broad, 1923; Cameron, 2015; Chisholm, 1981, 1990a, 1990b; Correia & Rosenkranz, 2018; Craig, 2000; Crisp, 2003, 2004; Forrest, 2005; Geach, 1972; Lowe, 1998, Ch. 4; Lucas, 1989; Markosian, 2004; McCall, 1994; Merricks, 1999; Prior, 1970, 2003; Tooley, 1997; Zimmerman, 1996, 1997, 1998.

distinction between what is past, present or future, while their opponents, the 'B-theorists' (or 'tense anti-realists')[3] deny the objectivity of any such distinction. To put it another way, A-theorists and B-theorists agree that every thing in time is 'past relative to' some things, 'future relative to' others, and 'present relative to' itself – just as every place on earth is south relative to some places, north relative to others, and at the same latitude as itself. But, whereas B-theorists regard this spatial analogy as deeply revelatory of the purely relative nature of the division of time, A-theorists claim that it is misleading. According to A-theorists, every thing in time that is past, present, or future in a merely relative sense is, in addition, past, present or future in a *non*-relative sense (i.e. past, present, or future *simpliciter*) (cf. Zimmerman, 2011: 163–164).

Although there is an ongoing debate about whether there is a genuine metaphysical conflict between the A- and the B-theory of time,[4] it is generally accepted that these two theories draw their legitimacy from different sources of evidence: whereas the A-theory seems backed by our intuitions, the B-theory is favored by the scientific community.[5] More specifically, the A-theorist generally emphasizes that her theory can account for *how* we ordinary think of time: (i) time has a direction,[6] (ii) 'our present' extends throughout the universe, (iii) the future is open whereas the past is fixed, etc. By contrast, the B-theorist generally emphasizes that her theory can account for what science, especially contemporary physics, says about time: (i) time has no direction (since there is no objective sense in which time is flowing one way rather than the other), (ii) there is no unique present (since there is no absolute relation of objective simultaneity), (iii) there is no asymmetry in openness between the future and the past (since the 'block universe' view of time is *isotropic* and the fundamental laws of physics are *time-reversal invariant*), etc.[7]

Of course, A-theorists disagree on many points (e.g., on whether all of reality is confined to the present), but they usually share the same high concern for accounting for our commonsense view of time, especially for our genuine conception of change. In that sense, the A-theorist's picture of the world is essentially *dynamic*:

[3] Frege, 1984 (see esp. p. 370); Grünbaum, 1967, Ch. 1; Le Poidevin, 1991; Lewis, 1986; Mellor, 1981, 1998; Quine, 1960, §36; Russell, 1903, Ch. 54; Saunders, 2002; Savitt, 2000; Sider, 2001; Smart, 1963, Ch. 7, 1987.

[4] Cf. Williams, 1996, 1998a, 1998b, 2003; Deng, 2010; Oaklander, 2001; Parsons, 2002.

[5] It should be noted, however, that this way of framing the debate between the A- and the B-theory is properly contemporary (and perhaps partly misleading). As one of my reviewers rightly pointed out, historically, the B-theory has often been motivated by theological considerations, or Platonism, whereas the A-theory has been motivated by scientific arguments (particularly associated with the life sciences).

[6] Even if no one (to my knowledge) has defended such a combination of views, it is possible to claim that time lacks an intrinsic direction but includes objective distinction between past, present, and future (cf. Sider, 2005).

[7] In this perspective, the A- and B-theory of time may both appear unsatisfying: the A-theory conflicts with science (especially with the Special and General theories of relativity), while the B-theory gives us no handle on time as universally experienced (especially in terms of an ongoing now) (cf. Baker, 2010: 27).

events are constantly changing with respect to their A-properties by first becoming less and less future, then becoming present, and subsequently becoming more and more past. The passage of time is thus depicted as a real and inexorable feature of the world. By contrast, B-theorists regard time as being closely akin to space, i.e. as a *static* dimension in which everything is *permanent*. After all, events do not change with respect to their B-relations: if the Battle of Waterloo is earlier than the conquest of Mars, then the Battle of Waterloo is *always* earlier than the conquest of Mars. The concern of A-theorists for change as we experience it plays a crucial role in McTaggart's argument for the unreality of time. This argument can be reconstructed in various ways (Kit Fine's version of McTaggart's argument will be examined below) but, as a first approach, it will suffice to consider the four following claims:

1. Genuine change is *essential* for the reality of time.
2. Only the A-series involves genuine change.
3. The A-series is either *contradictory*, or generates a *vicious circle* or an *infinite regress*.
4. Therefore, there is no genuine change, and so time is unreal.

First, McTaggart argues that genuine change is *essential* for the reality of time. He does not say much in support of this claim.[8] The reason is presumably that, since Aristotle (*Physics*, IV, 219a 4–6), time is most often defined as *the measurement of change*. In that sense, things change continually and we call 'time' the measurement, i.e. the counting of this change. This idea runs deep: time is what we refer when we ask 'When?'. 'After how much time will you return?' means 'When will you return?'. The answer to the question 'When?' refers to something that happens. For example, 'I will return in 2 days' means that between departure and return the sun will have completed 2 circuits in the sky (cf. Rovelli, 2018: 49).[9] So if nothing changes, time does not pass, because time is our way of situating ourselves in relation to what changes. If nothing changes, there is therefore no time. According to Aristotle, even when "[…] it is dark and our body experience is nil, but some change is happening within the mind, we immediately suppose that some time has passed as well" (1999: 105). In other words, even the time that we perceive flowing within us is the measure of a change.

We had to wait until the end of seventeenth century to get a completely different picture. In his *Principia*, Newton recognizes that there is a kind of 'Aristotelian' time that measures days and movements. But he also contends that, in addition to this, another time must exist: an 'absolute, true and mathematical' time that flows by itself, independently of things and of their changes. As Carlo Rovelli puts it: "[i]f all things remained motionless and even the movements of our souls were to be frozen, this time would continue to pass, according to Newton, unaffected and equal

---

[8] We find McTaggart accepting premise (1) in as early a work as his *Studies in the Hegelian Dialectic* ([1896] 2011: §5).

[9] In the Greek geocentric worldview, the planets move around the 'central' Earth with perfect regularity and invariance. That is precisely why Aristotle regards celestial motion as the best criterion of temporal passage.

to itself" (2018: 76). 'Absolute, true and mathematical' time, Newton says, is not directly accessible, only indirectly, through calculation. It is not the same as that given by days, because "[…] the natural days are truly unequal, though they are commonly consider'd as equal, and used for a measure of time: Astronomers correct this inequality for their more accurate deducing of the celestial motions" (Newton, 1729: 12). However, even if the Newtonian picture theoretically allows for 'temporal vacua' (i.e. periods of time during which nothing changes), it still seems true that time is the dimension in which changes *do* and *can* occur. So, McTaggart is certainly right in saying that a world without change is a world without time, i.e. a world in which change *cannot possibly* occur.

This latter claim has nonetheless been challenged by Sidney Shoemaker (1969) who has famously argued that time can pass without change. In particular, Shoemaker invites us to imagine a world divided into three regions: A, B, and C. From time to time, one or more of these regions get 'frozen' in the sense that change of all kinds comes to a complete halt within the region. When this occurs, the inhabitants of the frozen region observe nothing amiss: "[…] the period passes without their noticing it or measuring it in any way, so they rely on the testimony of the inhabitants of other regions in forming beliefs about their own frozen periods" (Le Poidevin, 2010: 172). Comparing their notes, the inhabitants of all three regions have found (by the use of clocks located in unfrozen regions) that: (i) each local freeze lasts for exactly 1 year and (ii) A is frozen every 3 years, B every four, and C every five. Since the three regions constitute the whole of this world, a simple calculation shows that, every 60 years, there is a *total* freeze lasting 1 year. Taking this scenario seriously,[10] the inhabitants of this world seem to have reasons "[…] for believing that there are intervals during which no changes occur anywhere" (Shoemaker, 1969: 371). However, although Shoemaker's argument is both imaginative and subtle, it misses its target. Many objections can be found in the literature (cf. Scott 1995; Warmbrod, 2017); the most striking one shows that Shoemaker begs the question: he assumes the possibility of changeless time instead of proving it. After all, as Denis Corish has pointed out, Shoemaker's scenario allows for another interpretation: "[…] instead of a year of changeless time passing, in say region A, there is no time there […]" (2009: 222). The only reason why Shoemaker does not retain this interpretation is that he *assumes* that "[…] if there is time in one place, there is the same time in another, distant, place, whether anything happening in that other place or not" (2009: 221). For instance, he assumes that the same year as passing in B occurs in A, though everything in A is frozen – this is the assumption of changeless time. But since this possibility is precisely what Shoemaker is arguing

[10] Some philosophers, e.g., David Hugh Mellor (1982) or J. L. H. Thomas (1991), argue that such fantasy scenarios supply no evidence for possibility at all, and should therefore not be considered. However, as Robin Le Poidevin has pointed out, this kind of objection cannot fairly be directed at Shoemaker, "[…] for his aim is not to argue directly for the logical possibility of time without change, but rather indirectly, by showing what is wrong with a peculiar kind of argument against that possibility" (2010: 173).

for, it cannot be assumed in the argument without begging the question. Also, there is change in McTaggart's A-series sense even during the frozen periods.

Second, McTaggart argues that only the A-series involves genuine change. This claim is more contentious than the first one. It rests on the idea that the only way in which events can genuinely change is by first being future, then present and finally past, i.e. by changing positions in the A-series. McTaggart thus denies that an event having different properties over time – such as a discussion that begins politely and becomes rude – is a paradigmatic example of change, since it is always the case that the earlier part of this event is politer than its later part. After all, supposing that the B-theorist is right and that time could exist without the A-series, everything is *permanent*: events possess different properties at different locations in a static dimension. So, just as variations in properties across space at a given moment of time do not amount to change (because all locations in question coexist), variations in properties over time do not amount to change either.[11] For this reason, McTaggart rejects Bertrand Russell's account of change, according to which something changes just in case a proposition is true at one time but not true when evaluated at a later time. As McTaggart insists, if a proposition p has some truth-value when evaluated at a time, it is *always* the case that p has *that* truth-value when evaluated at *that* time.[12]

The situation is quite different when introducing the A-series into the equation: we still have the permanent B-facts – the discussion is polite at t_1 and rude at t_2 – but in addition we have ever-changing A-facts. The discussion's being polite at t_1 is first in the distant future, then the near future, then the present and thereafter the ever more distant past. McTaggart hence concludes that genuine change requires ever-changing A-facts and, therefore, that an eternal B-world (i.e. a world without such facts) is a world where genuine change cannot occur and, in virtue of (1), is a world without time. To be sure, McTaggart does not disagree with the commonsensical view according to which a discussion that is polite at one time and rude at a later time has changed; his claim is rather that in the absence of ever-changing A-facts, there are no *times* at all, since change cannot occur.

Third, McTaggart argues that the A-series, which has been shown to be *essential* for the reality of time, entails a *contradiction*: every event in the A-series (assuming that there is no first or last event) has the mutually incompatible properties of *being past, being present* and *being future*. In that sense, since everything starts off as being future, then becomes present before sinking into the past, it follows that every event has all three A-properties, while no event can possess any two of them without

[11] As an example, consider a typical case of spatial variation: if a stick is red at one end and black at the other, this is not a change. Rather than one and the same entity (the whole stick) possessing different properties, we have two distinct parts of the stick possessing different properties – that is two numerically different entities (cf. Dainton, 2010: 39).

[12] In response, B-theorists can ask 'Why cannot some changeless facts *be* changes? If the discussion is polite at t_1 and rude at t_2, why cannot these facts not constitute a change in the discussion? Does it matter that these facts are and always will be facts?' McTaggart's response is to deny that B-theorists have any right to talk of *times* at all in the absence of A-change. However, since this is precisely the assumption that B-theorists reject, this response does not trouble them at all (Dainton, 2010: 38).

generating a contradiction. If an event is present, for example, it cannot also be either past or future. The obvious reply to this apparent contradiction is to say that no event has two or more of these incompatible properties *at the same time*, but rather has them successively at *different moments of time*. McTaggart is not blind to this obvious reply; as he puts it: "[i]t is never true, the answer will run, that *M is* present, past and future. It *is* present, *will be* past, and *has been* future. [...] The characteristics are only incompatible when they are simultaneous, and there is no contradiction to this in the fact that each term has all of them successively" (1908: 468).

However, McTaggart maintains that this obvious reply fails, because it involves either a *vicious circle* or *an infinite regress*. First, this reply is clearly *circular*: confronted with an apparent contradiction involving the A-series (every event possesses mutually incompatible A-properties), the A-theorist is employing the A-series itself to get around the problem. In particular, the A-theorist is claiming that every event in the A-series being past, present, and future, has these properties successively *at moments of time*, while the successive order of *those* moments of time needs to be described by invoking the A-properties themselves. This is not satisfactory: one cannot *presuppose* the A-series when trying to rid it of contradiction. Of course, as McTaggart concedes, a reply could be the following: "[o]ur ground for rejecting time [...] is that time cannot be explained without assuming time. But may this not prove – not that time is invalid, but rather that time is ultimate?" (1908: 470). For example, it may well be impossible to explain notions such as 'goodness' or 'truth' in completely different terms, but this does not lead us to reject the notions in question. So, perhaps time is also a primitive concept, one that cannot be explained in other terms. However, this reply does not seem to apply here: notions such as 'goodness' or 'truth', though they may not be explained in a non-circular way, do not involve any contradiction. The notion of time must therefore be rejected as paradoxical, unless the contradiction it involves can be removed.

In that respect, claiming that the contradiction inherent in the A-series can be removed by introducing a second-level A-series (*M* is present in the present, past in the future, and future in the past) does not help solve the problem, since this clearly involves an *infinite regress*. What then about the contradiction inherent in the second-level and following A-series? At any point in this regress at which we stop we are left with an A-series that suffers from the same difficulty as the previous one. For example, take an event such as the death of Kennedy and consider second-level properties such as *being present in the present*, *being past in the future*, *being future in the past*, etc. It clearly appears that some of *these* properties are incompatible: if the death of Kennedy is past in the present, it cannot also be either present in the present or future in the present. Of course, the A-theorist might again insist that this event does not have two or more of these second-level incompatible properties *at the same time*, but rather has them successively at *different moments of time*. But once again, the succession of *those* moments of time needs to be described by invoking third-level properties that will in turn be incompatible, so that the contradiction will never be removed by ascending in the hierarchy (no matter how many levels we add). This leads McTaggart to conclude that introducing higher-level properties fails to eradicate the initial paradox, since precisely the same paradox exists at each level of temporal predication. The A-series is therefore inherently paradoxical, and so it cannot correspond with anything in reality. Given this, McTaggart maintains that there is no change (in virtue of (2)) and, therefore, that time does not exist (in virtue of (1)). What does exist, according to McTaggart, is an atemporal 'C-series': "[…] an eternal four-dimensional block of events, whose contents are ordered but not in a temporal way" (Dainton, 2010: 17).

One may agree with Peter Geach that sometimes "[t]ime is so perplexing that we can understand philosophers' wishing to cut the knot by denying the reality of time […]" (1973: 213). In this perspective, McTaggart's argument is surely powerful, but suffers from at least two weaknesses: (i) it presupposes an ontology of events, and (ii) it neglects the variety of forms the A-theory may adopt in order to avoid the contradiction. That is partly *why* Kit Fine (2005: 272–273) proposes a revised version of this argument.[13] Specifically, Fine formulates four general metaphysical assumptions, the conflicts between which threaten the reality of time (cf. also Correia & Rosenkranz, 2012, Lipman, 2015: 3120–3121, Deng, 2017: 1116):

1. *Realism*: Reality is constituted (at least, in part) by tensed facts.[14]
2. *Neutrality*: No time is privileged, the tensed facts that constitute reality are not oriented towards one time as opposed to another.
3. *Absolutism*: The constitution of reality is an absolute matter, i.e. is not relative to a time or other form of temporal standpoint.
4. *Coherence*: Reality is not contradictory, it is not constituted by facts with incompatible content.

Fine's version of McTaggart's argument (at least in its simple form)[15] can be expressed as follows. *Realism* states that reality is constituted by some tensed facts. There is therefore some time t at which *this fact* obtains. Now, it follows from *Neutrality* that reality is not oriented towards one time as opposed to another. In particular, the totality of tensed facts constituting reality are not merely ones that *presently* obtain (the present is no privileged time). So, reality is presumably constituted by similar sorts of tensed facts that obtain at other times (the present is only

[13] Fine's version of McTaggart's argument differs from McTaggart's own version in at least two important ways. First, whereas McTaggart claims that the tense realist is required to hold that any given event is past, present, and future (cf. 1908: 468), Fine does not presuppose an ontology of events, nor does he suppose that the incompatibility lies in the determination of something as past, present, and future. Second, unlike McTaggart, Fine does not begin by supposing that "[…] there is a prima facie contradiction in the realist's position from which he must somehow extricate himself" (2005: 275). Rather, Fine's main goal is to demonstrate a contradiction, i.e. to spell out the assumptions and the reasoning by which the contradiction is derived.

[14] The assumption (1) expresses the commitment to tensed facts that tense anti-realists (or B-theorists) exclude.

[15] Fine presents a more sophisticated version of the McTaggart's argument, which does not take the notion of constitution as *basic* (since it indeed seems that there are various notions of constitution for which some of the four assumptions do not hold) (cf. 2005: 272).

one time among others, after all). Since any reasonable view of reality should allow for its being variegated over time, it must be assumed that the facts constituting reality that obtain at *t* are *incompatible* with the facts constituting reality that obtain at other times. For instance, if such a view allows for the present fact that 'Kit Fine is sitting', then it should also allow for the subsequent fact that 'Kit Fine is standing'. Finally, given *Absolutism*, reality is absolutely constituted by these facts, while this contradicts *Coherence* (cf. Fine, 2005: 272).[16]

The standard realist response to Fine's McTaggartian argument is to retain *Realism* but to reject *Neutrality* and, therefore, to claim that reality is somehow oriented. Fine calls those realists who reject *Neutrality* 'presentists'. Roughly, if there is no more than one time at which reality is constituted by tensed facts, which is intended to be the present time, then reality is not contradictory, since no two facts with incompatible content can *simultaneously* obtain. For example, reality is constituted by the present fact 'KF is sitting', but not by the subsequent fact 'KF is standing'. However, contrary to what Fine suggests, there are (at least) two other ways to reject *Neutrality*, i.e. to deny that the totality of facts constituting reality are those obtain at past, present and future times. Specifically, one can argue that (i) the totality of facts constituting reality are merely those obtained at past and present times (GBT), or (ii) the totality of facts constituting reality are merely those obtained at present and future times (Shrinking Block Theory – SBT).[17] It is worth noting that the moving spotlight (MSL) is not an option here, since it entails that the facts constituting reality are not oriented towards one time as opposed to another (no time is *ontologically* privileged) and, therefore, that *Neutrality* is retained. Although Fine only considers presentism, there are also various ways for the other two options (GBT and SBT) of avoiding the contradiction (cf. Correia & Rosenkranz, 2012, Broad, 1923: 79–84). This will later be a matter of great concern (cf. Sects. 3.9 and 3.10), but I can already give an insight into my own GBT solution to Fine's McTaggartian argument.

In order to derive a contradiction (e.g., reality is constituted by the facts 'KF is sitting' and 'KF is standing'), Fine *presupposes* that the tensed facts which constitute reality at more than one time do not undergo any intrinsic change as time goes by. For example, if the fact 'KF is sitting' constitutes reality at the present time, then the very same fact 'KF is sitting' constitutes reality at every future time (especially at every future time at which the fact 'KF is standing' also constitutes reality, which generates a contradiction). Yet, there might be a plausible alternative to this

[16] It is difficult to say whether this argument succeeds, since it is far from obvious how the four metaphysical assumptions are to be understood. As Correia & Rosenkranz have noticed, "[o]ne reason is that [these four metaphysical assumptions] involve non-orthodox talk of facts 'constituting reality'" (2012: 309). Perhaps the most straightforward way to understand the notion of constituting reality is to identify it with the notion of existing (as applied to facts).

[17] One must acknowledge that Fine's intention is distorted here, since he is explicit that talk of facts is not to be taken seriously in his paper: the official idiom is 'in reality, *p*', rather than 'the fact that *p* constitutes reality'. Nonetheless, taken literally, his presentation of the McTaggartian argument offers a powerful framework within which the A-theories of time can be expressed (though they are usually expressed in terms of which times, events, and things exist, rather than facts).

presupposition: whatever constitutes reality at the present time (I will speak of events rather than tensed facts) undergoes intrinsic changes by becoming past, to such an extent that it does no longer belong to its natural kind. This change will be introduced as an alteration of certain intrinsic properties, namely the properties that make, for instance, an event belong to the natural kind to which it belongs when present (rejection of *natural kind essentialism*). Just as a bronze statue exists in a changed form after having been melted, an event exists in a changed form after having become past. For example, the battle of Waterloo is an event when occurring at the present time, but is no longer an event when past (without it ceasing to exist). Once this alternative is accepted, Fine's argument for the joint inconsistency of non-presentist rejections of *Neutrality*, *Realism*, *Absolutism*, and *Coherence* is blocked, since reality is not constituted by the very same entities (tensed facts, events, etc.) at present and past times. For instance, the fact 'KF is sitting' constituted reality one minute ago, when he was sitting, while it is no longer a fact *now*, when he is standing. Provided that the fact (the event) 'KF is sitting' still exists in a changed form, this solution seems compatible with the claim that nothing ever ceases to exist. There therefore are no times at which two same entities, with incompatible content, are *both* constituting reality.

The non-standard realist response to Fine's McTaggartian argument is to retain *Realism* but to reject either *Absolutism* (relativism – reality is constituted by all the tensed facts that obtain at some time or other, but it is constituted by them at different times)[18] or *Coherence* (fragmentalism – reality is constituted by all the tensed facts that obtain at some time or other, even though some of these are incompatible). These two non-standard responses, Fine claims, are the best options for tense realists to account (i) for the passage of time, (ii) for the connection between language and reality, and (iii) for the compatibility between tense realism and special relativity (cf. 2005: §§7–9). However, since the non-standard options are not fully intelligible, they will not be considered in the remainder of this chapter. Specifically, one reason why *relativism* is not fully intelligible is that, contrary to what this theory implies, temporal reality is naturally thought as being 'one' (rather than 'many'): "[j]ust as there is not one reality where you are and another where we are, there does not seem to be a succession of distinct realities corresponding to different times [...]" (Correia & Rosenkranz, 2012: 311). Admittedly, from an historical point of view, it sometimes makes sense to speak of different epochs (e.g., the Enlightenment, the Victorian era, etc.) as corresponding to a succession of realities. This allows one to insist on the political, economic, and sociocultural differences between epochs. But, "[...] when it comes to metaphysics, one tends to see these different epochs as parts of a single reality" (Correia & Rosenkranz *id.*). Thus, since *relativism* – that

[18] Of course, the standard realist can admit a sense in which the constitution of reality might be relative (e.g., by saying that for reality to be constituted by a tensed fact f at t is for it to be constituted by f whenever t is present). However, the non-standard realist takes reality to be relative in a more radical sense. For him, reality is *irreducibly* relative: "[...] reality at another time is an *alternative* reality [...] on an equal footing with the current reality" (2005: 279).

views temporal reality as a mere collection of different successive realities – goes against this unified picture, it is hard to give it metaphysical relevance.

Then, one reason why *fragmentalism* is not fully intelligible has to do with the very notion of 'coherence' at work in the definition of a fragment – a fragment is a maximal collection of mutually coherent facts. According to *fragmentalism*, temporal reality is 'one' (i.e. there is no diachronic shifts in what facts constitute it), but is somehow 'fragmented', insofar as it is constituted by facts with incompatible content across the fragments. Now, the notion of 'coherence' at work in the definition of a fragment – call it 'coherence*' – cannot be the ordinary notion of coherence that holds amongst propositions. Otherwise, "[...] the reduction of times to fragments would be inadequate due to an over-generation of fragments. We would find more times than there actually are" (Lipman, 2015: 3124). For example, perhaps there is no time at which two facts – e.g., 'Socrates is furious' and 'Plato is anxious' – obtain, so that a fragmentalist should say that there is no fragment that comprises both of these facts. Yet, these two facts clearly cohere: "[...] it could have been the case that, at some time, Socrates is furious and Plato is anxious" (Correia & Rosenkranz, 2012: 312). Although Fine suggests that the notion of coherence* should be taken as *primitive* (2005: 281), this notion is too mysterious for this being an option. Thus, since it is far from clear how the crucial notion of 'coherence*' should be understood, it is hard to make sense of *fragmentalism*. Both standard and non-standard positions are summarized in Fig. 3.1.

Thus, although McTaggart's argument takes the moving-spotlight theory (which results from the acceptance of the four Finean assumptions and, therefore, leads to contradiction) as its primary target, Fine's revisited version of the argument shows that other sorts of tense realism (both standard and non-standard) can be identified. In general, since tense realists disagree about how *Neutrality* should be rejected (e.g., is all of reality confined to the present?), it is common to define the different versions of the standard A-theory by how they answer to the question 'Do the future

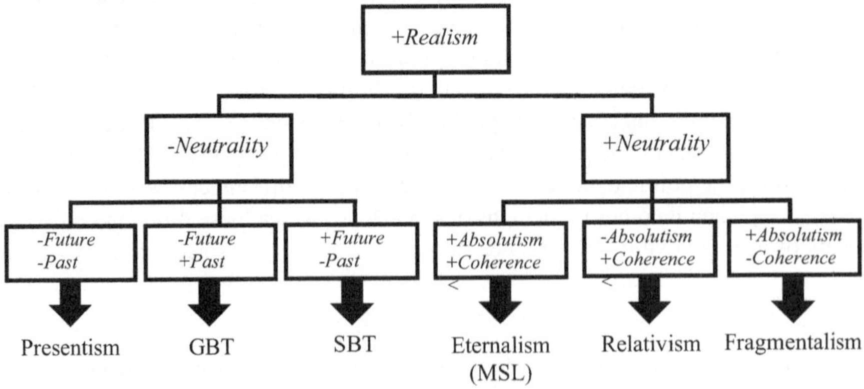

Fig. 3.1 Standard and non-standard realist theories of time

and the past exist?'.[19] According to the traditional definitions, *eternalism* and *presentism* are the two extreme answers that can be provided to that question: the former says that both the future and the past exist, whereas the latter says that neither the future nor the past exists. Intermediate between these two extreme answers are the *growing block theory* (GBT), which says that the past exists but the future does not exist (cf. Sider, 2001: 12, Callender, 2011: 3), and the *shrinking block theory* (SBT), which says that the future exists but the past does not exist. For example, as Caspar Hare puts it: "[s]ome imagine that the past exists but the future does not [...]. Some imagine that the future exists but the past does not [...]. Presentists, meanwhile, hold that only present objects, events, moments exist (and perhaps things that exist timelessly, like gods and numbers). There are no past or future things" (2009: 17).

However, there are some reasons to complain about this traditional way of defining the A-theories (cf. Williamson, 2013: 24, Deasy, 2017: 381–389). In particular, it leads to think of GBT as a hybrid between eternalism and presentism, with the consequence that this theory is often criticized for accumulating the flaws that are identified in the two extreme forms of the A-theory. For example, just like eternalism (at least in its A-theoretic version), GBT seems to face with the so-called 'epistemic objection' (according to which there is no way of knowing that our time is the objective present) and, just like presentism, GBT seems to conflict with the theories of relativity (e.g., by requiring an absolute notion of objective simultaneity). These two issues will be a matter of great concern in the remainder of the book. Specifically, in the fifth section of this chapter, I will argue that the 'epistemic objection' relies on a mistaken assumption, namely that past beings (e.g., Cesar, Napoleon) are still believing to be in the objective present, and therefore that it does not apply to GBT. Further, in the next chapter, I will argue that GBT, far from being disqualified by contemporary physics, might be underpinned by some recent approaches to quantum gravity. As a preliminary step, however, it will be useful to change the way GBT is generally perceived among the A-theories of time in order to reveal the whole potential of this theory (especially when it comes to accommodate the asymmetry between the open future and the fixed past). In the following sections, I will therefore first express my dissatisfaction regarding the traditional way of defining the A-theories of time, and then make a fresh start in the debate by introducing two questions whose specific answers will allow (i) to get a comprehensive categorization of the A-theories of time and (ii) to reveal the metaphysical singularity and potential of GBT.

[19] This way of putting the question purposely remains general: 'the future' and 'the past' not only refer to instants, but also to events (e.g., WWI, the conquest of Mars) and things (e.g., Napoleon, my great grandson).

3.3 Complaining About the McTaggartian Picture

One may agree with Timothy Williamson (2013: 22) that there is a "feeling of dissatisfaction" with the eternalism-presentism distinction. The reason, he says, is that there is no satisfactory way to spell out what is meant by 'is present' in the traditional definition of presentism, according to which 'everything is present'.[20] There are indeed good reasons to reject all the most plausible candidate interpretations. For example, consider the natural idea that "[…] to be present is just to be real or to exist" (Zimmerman, 1996: 117). The problem with this interpretation is that it makes presentism trivially true: if presentism is to be interpreted as the thesis that 'everything exists', then everyone is a presentist! Another natural idea is to claim that to be present is to instantiate the primitive property of presentness. However, as Zimmerman himself claims, "[…] no *real* presentist has any reason to believe in a special quality of 'being present'" (1996: 125). Moreover, as Daniel Deasy argues, this interpretation reduces the debate between A-theorists to the debate about whether everything or only something instantiates this property, which sounds like "[…] a parody of the philosophy of time" (2017: 383). Finally, Williamson mentions a third interpretation: "[…] something is present when and only when it is spatially located" (2013: 24). However, it is not clear that this interpretation would be false in a non-presentist setting (as it is meant to be) – there would, for instance, be nothing *prima facie* problematic in an eternalist's asserting that everything has a spatial location. Furthermore, this latter interpretation makes presentism incompatible with theories that have no ramification with the philosophy of time, such as the platonist theory that there are spatially unlocated abstract objects (e.g., numbers). Since the predicates 'is past' and 'is future' – as they appear in the other A-theories of time (MSL, GBT and SBT) – are respectively defined in terms of 'to be *earlier* than the present things' and 'to be *later* than the present things', the mystery surrounding the term 'is present' seems to infect all the traditional definitions.[21]

Another issue raised by the traditional definitions of the A-theories of time has to do with the interpretation of the universal quantifier 'everything' (cf. Crisp, 2004; Ludlow, 2004; Meyer, 2005, 2011; Miller, 2013; Sider, 2006a). The question is whether, when presentists claim that 'everything is present', the quantifier is to be interpreted as tensed or tenseless. Both options seem problematic. If it is tensed, then presentism is the trivially true thesis that 'everything present is present' (which hence makes eternalism trivially false). If it is tenseless, then presentism is the trivially false thesis that 'everything past, present or future is present' (which hence makes eternalism trivially true). It therefore seems that presentism is either trivially

[20] As Thomas Crisp makes clear, "[t]his way of putting the thesis, or something close to it, is fairly common in the literature" (2004: 15). For instance, Merricks characterizes presentism as the view that "all that exists, exists at the present time" (1995: 523), and Bigelow as the view that "nothing exists which is not present" (1996: 45).

[21] Williamson therefore proposes to abandon the eternalism-presentism distinction as "hopelessly muddled", and to get on with the clearer permanentism-temporaryism debate (cf. 2013: 25).

true (and eternalism trivially false), or trivially false (and eternalism trivially true). In other words, it appears that the traditional way of defining the A-theories of time turns what is meant to be a metaphysical debate between two venerable positions – presentism and eternalism – into a purely semantic debate about *how* the universal quantifier in the definitions should be interpreted. This is what leads Ulrich Meyer, for instance, to conclude that "[…] there is no reading on which [presentism] expresses a substantial metaphysical truth" (2005: 213–214).

However, although it should be acknowledged that traditional definitions of the A-theories of time (i) make unclear what it is for something to be present, and (ii) threat the metaphysical nature of the eternalism-presentism debate, this is not what should concern us most. After all, following Williamson (2013), Correia and Rosenkranz (2018) have shown that presentism (and other A-theories) can be defined without using the mysterious notion of 'presentness'. In particular, they argue that presentism is the only theory that accepts both of the following principles: (P2) 'every time is new at itself' ($Tx \rightarrow At\, x, H\neg E!x$), where 'T' is a predicate for times and 'H' stands for 'Always in the past', and (P3) 'every time is last at itself' ($Tx \rightarrow At\, x, G\neg E!x$), where 'G' stands for 'Always in the future'.[22] Furthermore, it is not clear that in using 'everything' in the tenseless way, presentists are saying something trivially false. As Correia and Rosenkranz (2015, 2018) have argued: just as saying that 'every black or non-black raven is black' is "[…] a perfectly sound way of saying that the only ravens that exist are black", saying that 'everything past, present or future is present' is "[…] a perfectly sound way of saying that the only things in time that exist are present – and this will remain to be so even if it is assumed that what is present may also be past or future" (2018: 62). For his part, Deasy (2017: 381) argues that there is a natural reading of the traditional definition of presentism on which it expresses a thesis which is neither trivially true nor trivially false: '**A** $\forall x$ *Present*(x)', where '**A**' stays for 'always' and '$\forall x$' is the universal quantifier of classical first-order logic and is, following Marcus (1962), neither tenseless nor tensed.

Rather, the main problem with the traditional way of defining the A-theories of time is that it rests on an ontological question – 'Do the future and the past exist?' – while we have no strong intuitions thereon. A-theories of time are indeed meant to match the ordinary intuitions we have regarding the nature of time (e.g., time flows uniformly and universally, our 'present' extends throughout the universe, the future is open while the past is fixed, etc.), and if we define these theories in terms of whether the future and the past exist, then we risk simply missing the point. For instance, a statement such as ⌜The conquest of Mars exists⌝ can neither be confirmed nor refuted by any intuition or experience whatsoever. As Clifford Williams puts it: "[…] there is no experiential way to differentiate between [events] being equally

[22] Another possibility is Daniel Deasy (2017), who claims that presentism should be identified with the only A-theory which accepts *Transientism*, i.e. the principle that 'Sometimes, something begins to exist and sometimes, something ceases to exist'. These strategies (involving presentism without presentness) will be a matter of great concern in the next chapter (Sect. 4.4), since they might offer a solution to the Putnam-Rietdijk argument against the A-theory of time.

real and not being equally real" (1998b: 386); the only events we experience are the ones occurring at the time of our experience, whether this be in eternalist or in presentist settings. For his part, Craig Callender compares the eternalism-presentism debate to two people arguing about whether the refrigerator lightbulb goes out when the door is shut: "'[R]efrigerator presentists' believe the light is off when the door is shut; 'refrigerator eternalists' believe the light remains on" (2000: 588). This debate is pointless since, barring drilling a hole into the side of the refrigerator, we can only check the light by opening the door (while this will not test either hypothesis).[23]

A reply might be that empirical science, especially observational astronomy, is precisely what allows us to drill a hole into the side of the refrigerator and, therefore, to determine *who* of the eternalists or the presentists are right. For example, it might seem that the recent observation of gravitational waves speaks in favor of eternalists, since this observation informs us about the very early universe. But, as it should be clear, all that the observation of gravitational waves allows us to conclude is that the past existed (not that the past is still existing), while this is compatible with both eternalist and presentist theories. When we observe gravitational waves, we do not observe the past (or do so only in a metaphorical sense); we observe disturbances in the curvature of spacetime that, although they emerge from the very early universe (and might therefore offer a unique probe to explore it), are *simultaneous* with our observation. So, even empirical science – with its powerful instruments and super sensitive detectors – only allows us to observe events that occur at the time of our observation. It therefore seems hopeless to appeal to empirical data in order to settle the debate between eternalists and presentists.

To the question 'Do the future and the past exist?', we thus have no pre-theoretic answer. The dispute between eternalists and presentists cannot be solved by means of intuition – that is probably *why* this dispute may appear so stipulative (cf. Williams 1998a, 1998b; Callender, 2000; Dorato, 2006a; Savitt, 2006). Of course, this does not mean that it is *false* to say that the various A-theories of time differ with respect to how they answer to the question 'Do the future and the past exist?'. For instance, it is usually *right* to say that eternalists hold that both the future and the past exist, while presentists hold that neither the future nor the past exist. What I ultimately want to argue is that, since these definitions can neither be confirmed nor refuted pre-theoretically, they cannot be *essential* to the A-theories, i.e. they cannot be true in virtue of *the nature* of the A-theories, which are primarily meant to match our ordinary intuitions on time. Of course, one could object that matching (at least some of) our intuitions should not be A-theorists' primary concern; but, if so, then it is not clear *why* one should not adopt a B-theory of time which, as has been said, has the advantage of being favored by the scientific community. As a

[23] There is, of course, a sense in which we do experience earlier and later events: when we were at past times, we experienced the events occurring then, and when we will arrive at future times, we will experience the events occurring then. But this is true in both eternalist and presentist settings, so that this does not provide a clear-cut answer to the question as to whether the future and the past exist (cf. Williams 1998a, 1998b: 386).

reminder, the theories of relativity seem to imply a static view of time, in which past, present and future are equally real (the 'block universe' view of time) (cf. Sect. 3.2).

This approach could be criticized for ascribing more weight to intuitions than they can actually bear. After all, the controversy between A-theorists of time is primarily a metaphysical one, not an experiential one. The answer, it seems to me, is that there are several kinds of metaphysical controversies. Admittedly, many metaphysical controversies cannot be solved solely by appeal to experience. For example, if one wants to challenge Williamson's claim that vagueness is a form of ignorance (cf. Williamson, 1994), one will not deny that we experience some borderline cases; it is something that both Williamson and his critic agree upon.[24] What one will criticize is rather Williamson's inference from this experience, namely that there are sharp boundaries but we are unable to figure out their exact location. However, there are also metaphysical controversies that are more closely connected to experience. For example, Whitehead ([1934] 2011) argues that if we look carefully at our experience, we will find that fundamental concepts are not Aristotelian substances but activity and process: temporal entities are what we experience as basic, not concrete objects (which are regarded to be composites of many occasions of experience). If Whitehead is right, then the conflict between these two ontologies is "[…] decidable more by probing our experience than by making inferences from them" (Williams, 1998b: 391).

The debate on the nature of time seems more like the second of these disputes than the first. In the first, we all agree that we experience borderline cases, but we do not appeal to any experience to state whether these cases are ontic, semantic, or epistemic phenomena. It is the reverse with respect to the debate on the nature of time. Eternalism and presentism are not to be conceived as rival metaphysical explanations of one commonly agreed upon set of experiences. They are indeed not inferred from a prior knowledge of the temporal structure of the world. Rather, the debate on the nature of time is (partially) about which theory experience confirms. For instance, it is common for A-theorists of time to argue that a particular version of the A-theory is true partly because it matches with *how* time is ordinary experienced (see, for example, Zimmerman 2008: §7). Experience is thus evidence for theories of time, not the other way around. Therefore, one can reject the traditional way of defining the A-theories of time on the ground that the definitions can neither be confirmed nor refuted by experience, without jeopardizing the metaphysical nature of the debate. Some metaphysical debates are much more closely connected to experience than others, and the debate on the nature of time is presumably one of them.

A second, though less important, reason why traditional definitions should not be taken as *essential* to the A-theories is that there are some specific times at which they fail to distinguish between the theories in question. For example, at the start of

[24] I leave aside the nihilist position according to which vague terms (e.g., 'bald', 'child', etc.) cannot apply to anything (cf. Unger, 1979).

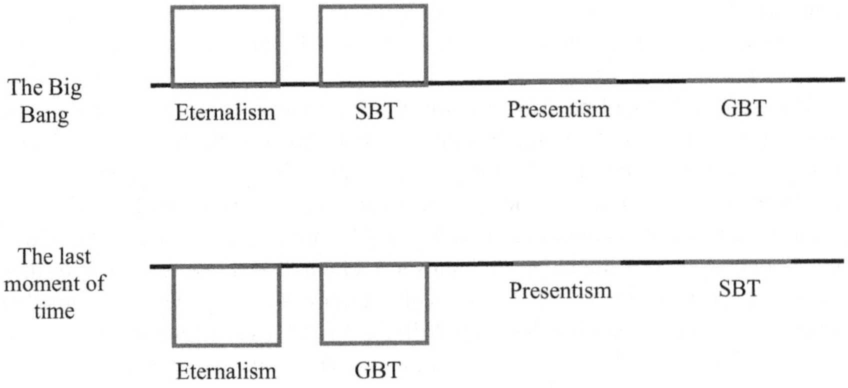

Fig. 3.2 A-theoretic structures at first and last moments of time

the Big Bang (where there is no past), eternalism and SBT, on the one hand, and presentism and GBT, on the other hand, are ontologically indiscernible: eternalists and shrinking blockers are both merely committed to the existence of the present and the future ('everything is either present or future'), whereas presentists and growing blockers are both merely committed to the existence of the present ('everything is present'). Likewise, at the last moment of time (where there is no future), eternalism and GBT, on the one hand, and presentism and SBT, on the other hand, are ontologically indiscernible: eternalists and growing blockers are both merely committed to the existence of the past ('everything is past'), whereas presentist and shrinking blockers are both merely committed to the existence of the present ('everything is present') (cf. Fig. 3.2).[25] Therefore, since the traditional definitions of the A-theories do not allow us to distinguish between them *at all times*, it seems that they cannot be considered as *essential* to them. After all, the following principle looks plausible: if there is at least one time *t* at which the definition *D* fails to singularize *x*, then *D* is not true in virtue of *x*'s nature, i.e. *D* is not *essential* to *x*.

Thus, the fact that the traditional way of defining the A-theories of time (i) neglects the role that intuitions are meant to play in the debate on the nature of time, and (ii) fails to distinguish between the A-theories *at all times*, suggests that the A-theories should be introduced differently. In the next section, I therefore propose a fresh way of singularizing the A-theories of time, by introducing two questions – 'Is there a geometric asymmetry between the future and the past?' and 'Is temporal becoming (i.e. the creation of new things in the present) real?' – the respective answers of which can be pre-theoretically evaluated. When these answers are superimposed, they allow one to get a comprehensive categorization of the A-theories of time. One immediate consequence of this proposal is that GBT will no longer represent a hybrid between eternalism and presentism, but a theory in its own right,

[25] Admittedly, this second objection does not succeed if the A-theories of time are defined as claims of type 'Always, …'. For, if presentism says 'Always, what exists is what is present', then, even at the start of the Big Bang, presentism and GBT are distinguishable.

which seems better designed than its rivals to accommodate the intuitive asymmetry between the open future and the fixed past.

Two concessions must nevertheless be made here before we can proceed. First, although GBT and SBT are marginalized by the traditional way of framing the A-theories of time, they have been defended in important and recent publications. Concerning GBT, one can mention Briggs and Forbes (2012, 2017, 2019), Deng (2017), and Correia and Rosenkranz (2013, 2018). Concerning SBT, publications are rarer, but one can still mention Casati and Torrengo (2011) and Norton (2015). Second, the ontological way of framing the A-theories of time, in terms of whether the future and the past exist, does not fully reflect the richness of the current debate. Many sophisticated views, such as McCall's (1994) shrinking tree, have been widely recognized as important features of temporal ontology. In that sense, although the eternalism-presentism distinction might appear dominant, it should not overshadow the fact that many philosophers have already proposed to reformulate the debate in order to capture the nuances it contains – some of these proposals will be discussed in the next section. The goal of reframing the debate is therefore not original in itself, but the reasons for which this reframing is undertaken in this book and the reframing itself are.

3.4 The McTaggartian Picture Revisited

As has been said, we have no intuition as to whether the past and the future exist. But an intuition that we surely have regarding the nature of time is that the future *differs* from the past. For example, causes occur prior to their effects, the arrow of time points from past to future, the future is open while the past is fixed, etc. In the previous chapter, I provided various reasons to think that this latter manifestation of *how* the future differs from the past should be explained by *how* reality truly is. It might therefore be tempting to extend this result to a large range of intuitive past-future time differences. John Earman (1974), for example, seems to yield to the temptation, when he argues for *The Time Direction Heresy*, i.e. the thesis according to which 'temporal orientation' (understood as a thicket of differences between the future and the past) cannot be reduced to non-temporal features.[26] Likewise, Tim Maudlin (2002, 2007) argues that a wide variety of time-asymmetries (including the passage of time and the direction of time) are ultimately grounded in the intrinsic time-asymmetry of the universe.[27] A third example is George Ellis (2006, 2013), who argues that the various arrows of time (e.g., the passage of time, the openness of the future) derive from an evolving block universe. All these approaches take our

[26] To be precise, Earman does not unequivocally endorse *The Time Direction Heresy*, but argues that no convincing arguments against it could be found in the very extensive literature.

[27] As will be explained below, I agree with Earman and Maudlin's general proposal according to which the universe encodes an intrinsic time-asymmetry. In particular, like Maudlin, I deny any attempt to reduce the time-asymmetry to the increase of entropy (Sect. 2.6). But I differ with respect to how the intrinsic time-asymmetry is established and what it is.

intuition that the future differs from the past *seriously*, i.e. as reflecting the deep structure of reality. To put it another way, all these approaches bridge the gap between *phenomenal* and *physical* time, by taking some of our intuitions as reliable indicators of how reality truly is.

Of course, one might object that, without further arguments (of the sort provided in the previous chapter), it is not clear that a large range of intuitive past-future time differences need to be grounded in the structure of reality. For example, it might be argued that causal direction is due to a projection of our experience as agents (Price, 1996, 2007), or to *The Past Hypothesis* (Albert, 2000, 2015).[28] Each of these claims provides an explanation as to *why* causes occur prior to their effects, without appealing to the deep structure of reality. The answer, it seems to me, is that as soon as one accepts that an intuition (e.g., the intuition that the future is open while the past is fixed) should be explained by the intrinsic structure of reality, all the alternative accounts for similar intuitive past-future time differences become *redundant*. This is not to say that all alternative accounts for temporal differences are false, but rather that they are *non-fundamental*, in the sense that they can ultimately be explained by the asymmetric nature of time itself. In other words, local phenomena (e.g., entropy, irreversibility) may not be as important as some philosophers would have us believe for past-future time differences. A proposal would therefore be to abandon the traditional debate (since there are no pre-theoretic reasons to adopt either eternalism or presentism), and to make a fresh start with a clearer distinction between two groups of A-theories of time: those that have intrinsic resources to explain *why* (at least some) differences between the future and the past are intuited, and those that need extrinsic resources.

This proposition immediately raises a question: 'What does 'intrinsic' mean here?' Roughly, an A-theory has intrinsic resources to explain *why* (at least some) differences between the future and the past are intuited iff the spatiotemporal structure it describes encodes a relevant past-future time asymmetry in its geometric features. To understand this, it is crucial to reconceptualize the A-theories as primarily describing contentless geometric constructions, whose features do not refer to objects antecedently given (by some sort of experience or prior knowledge) but only have a purely 'formal-logical' meaning stipulated by some primitive axioms (which therefore serve as implicit definitions).[29] In that sense, A-theories of time are not empirical constructions, but result from irreducibly theoretical choices: they

[28] According to *the Past Hypothesis*, the universe was, at one time in the very distant past, in a state of very low entropy. When combined with (i) the Newtonian Laws of Motion and (ii) a probability postulate, *the Past Hypothesis* explains our causal inferences from past to future (cf. also Ismael, 2016: 141).

[29] This 'conventionalist' geometry (which essentially makes no reference whatsoever to any kind of extra-formal content) should not be confused with applied (or physical) geometry, which attempts to "[…] coordinate such uninterpreted formal system with some domain of physical facts given by experience" (Friedman, 2002: 121). For example, according to the general theory of relativity, the geometry of physical space is another physical field (the field mediating specifically gravitational interactions): "[w]hether a given region of physical space is Euclidean or non-Euclidean depends on the distribution of matter and energy" (Friedman, 2002: 122). That is what led the physicist H. P. Robertson to famously conclude that geometry has become a branch of physics (cf. 1949: 315–332).

primarily outline geometric structures in which individuals (e.g., spacetime points, events, and objects) participate. Of course, this does not prevent A-theories from being subsequently confirmed (or refuted) by experience; it is even a crucial step! But, as has been said, experience is to be understood as evidence for the A-theories of time, not the other way around (cf. Sect. 3.3).

There are reasons to believe that the geometric structures described by the A-theories are *fundamental*, in the sense of not being dependent on sets of spacetime points, events or objects. Against a powerful tradition, which takes spacetime geometry to be a system of external relations that are instantiated by spacetime points, the idea is that spatiotemporal structures are ontologically primary, while individuals (such as spacetime points, events, and objects) have a mere derivative status. This 'structuralist' view is directly inspired from an influent current in the philosophy of science, sometimes called Ontic Structural Realism (OSR), which inflates the ontological priority of structure and relations.[30] However, whereas OSR is generally motivated by the interpretation of quantum physics (cf. Ladyman & Ross, 2007), the present structuralist view is motivated by the nature of the A-theories itself. Once again, A-theories are not set up from objects antecedently given; they primarily outline geometric background structures, the properties of which are stipulated by some primitive axioms. Therefore, taking A-theories seriously (i.e. as saying something about *how* time truly is), it is natural to regard the background structures they outline as fundamental and, thus, as ontologically prior to the individuals that participate within them. As a result, although some important geometric features of the spatiotemporal structures can hardly be expressed with no reference to individuals (such as spacetime points, events, and objects), one has to keep in mind that these individuals only play a *heuristic* role: they allow for the description of geometric structures which then carry the ontological weight.

Given this 'geometric' reconceptualization of the A-theories of time, 'the present' can be regarded as an axis around which some transformations can be operated. In particular, in the Euclidean plane, reflection symmetry is known as a transformation that preserves all geometric features. Such a transformation can, for instance, be operated on spatiotemporal structures, understood as primitive systems of fundamental spatiotemporal relations and derivative spacetime points. Interestingly, when reflection symmetry is operated around 'the present' axis of a spatiotemporal structure, the outcome is either an unchanged (or invariant) structure (when it is described by eternalism or presentism), or a transformed structure (when it is described by GBT or SBT). The reason is roughly that, unlike eternalism and presentism, GBT and SBT do not take what lies below (or above, respectively) 'the present' axis (viz. spatiotemporally related points) to be a structural reflection of what lies above (or below) it (cf. Figure 3.3). Therefore, whereas the structures described by eternalism

[30] OSR admits at least two forms: (i) *radical* – there are no individuals, but there is a relational structure (French & Ladyman, 2003; Ladyman, 1998), (ii) *moderate* – there are relations that do not depend on the intrinsic properties of their *relata* (Esfeld, 2004; Esfeld & Lam, 2008). However, the radical form of OSR is often criticized on the grounds that there cannot be relations without *relata* (cf. Busch, 2003; Cao, 2003; Morganti, 2004; Psillos, 2001, 2006).

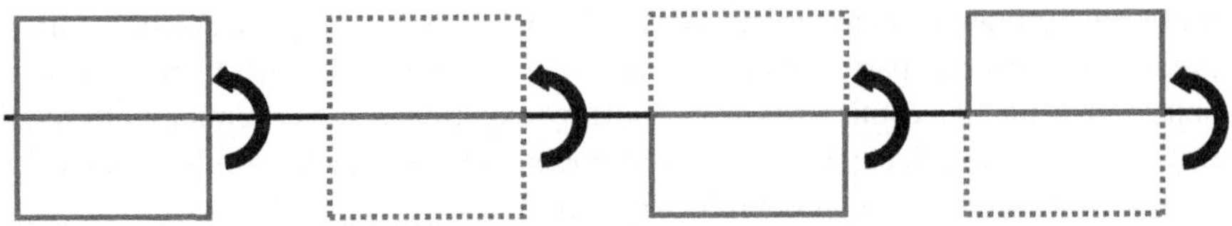

Fig. 3.3 Reflection symmetry on A-theoretic structures

and presentism do not change upon undergoing a reflection symmetry around 'the present' axis, those described by GBT and SBT do change: by 'symmetrizing' the growing block structure one obtains the shrinking block structure, and by 'symmetrizing' the shrinking block structure, one obtains the growing block structure. It thus seems that a simple operation, such as reflection symmetry, when applied around 'the present' axis on spatiotemporal structures, allows one to distinguish between two groups of A-theories of time: the *symmetric* theories (e.g., eternalism, presentism) and the *asymmetric* theories (e.g., GBT, SBT). As a first approach, an A-theory is called 'symmetric' if, when reflection symmetry is applied through 'the present' axis on the spatiotemporal structure it describes, we are left with an unchanged (or invariant) structure. Conversely, a theory is called 'asymmetric' if, when reflection symmetry is applied through 'the present' axis on the spatiotemporal structure it describes, we are left with a transformed structure.

However, it might be objected that there are versions of eternalism for which reflection symmetry around 'the present' axis fails to deliver an invariant structure: (a) if time has a beginning but no end (or vice versa), and (b) if time has both a beginning and an end, but the present is not equidistant from them. For instance, take any time t in version (a), reflection symmetry around t does change the structure; it delivers a structure where time has no beginning but an end (or vice versa). Of course, it could be replied that this is no big deal, since the opposition between symmetric and asymmetric theories is only meant to carry out a first sorting: symmetric theories include classical forms of presentism and eternalism, whereas asymmetric theories include classical forms of GBT and SBT. In that sense, this opposition does not rule out the possibility of developing non-classical forms of eternalism, which do not satisfy the geometric criterion, provided that they can subsequently be distinguished from both GBT and SBT. That is fair enough, but a better reply would be to define symmetric and asymmetric theories admitting no exceptions. In this regard, appealing to a modal characterization of these theories[31] could be salutary: a symmetric theory is a theory such that possibly always the structures it describes is reflection invariant; conversely, an asymmetric theory is a theory such that necessarily sometimes the structure it describes is not reflection invariant. The qualification 'sometimes' is there because (i) in a given world at the first moment of time (if there is one) the structure described by GBT is reflection invariant, and (ii) in a given

[31] This idea was suggested to me by Fabrice Correia.

world at the last moment of time (if there is one) the structure described by SBT is reflection invariant. Given this modal characterization, non-classical forms of eternalism (a) and (b) also fall under the scope of 'symmetric theories'. By convention, 'eternalism' will refer to eternalism *tout court* (i.e. classical or not) in the remainder of the text.

Since spatiotemporal structures have no more fundamental features than geometric ones (these features are theoretically posited, after all), *asymmetric* theories (e.g., GBT, SBT) are naturally seen as better suited than *symmetric* ones (e.g., eternalism, presentism) to account for (at least) some intuitive differences between the future and the past. Whereas *asymmetric* theories provide, through the geometric features of the structures they outline, an immediate, fundamental, and relevant reason as to *why* the future intuitively differs from the past, proponents of the *symmetric* theories must provide further explanations, presumably involving local phenomena (e.g., entropy, irreversibility), to account for the same intuition. In particular, the asymmetry is called 'relevant' because, as will be made clear in Sect. 3.5, it seems required to accommodate, for example, the 'no fact of the matter' account of openness introduced in the previous chapter (Sect. 2.9). Of course, this does not mean that if one accepts an *asymmetric* theory of time, one has to invoke geometric features to explain one or another intuitive past-future time difference. As Ross Cameron (2015: 194) makes clear, a growing blocker can perfectly claim that the geometric structure she describes plays no role in the way time is commonly intuited. But, it seems that the most natural move for her is to take widespread intuitions of *how* the future may differ from the past as manifestations of the intrinsic time-asymmetry of her model.

However, it might be objected that this use of geometry is analyzable in ontological terms and, therefore, that the revisited picture does not genuinely differ from the traditional one. For instance, it might be claimed that to say that the past geometrically differs from the future is only a sophisticated way to say that the past exists, while the future does not. This would clearly be problematic, since the main reason *why* the traditional picture was abandoned is that one has no strong intuitions on whether the future or the past exists. As a reminder, there is no experiential way to differentiate between the past and the future being equally real and not being equally real. However, there are (at least) 3 reasons to think that this objection does not apply here. First, one can agree on the asymmetric nature of time, while disagreeing on what exists. For example, growing and shrinking blockers agree that necessarily sometimes the future geometrically differs from the past, while they always disagree about what exists (either both the present and the past, or both the present and the future). Second, it clearly appears that a theory such as 'forward-branching time' makes a geometric distinction between the future and the past (the future is branching, the past is singular), while this theory rests on an eternalist ontology (where both the future and the past exist) (cf. Sect. 2.8). Finally, although it should be acknowledged that reflection symmetry on spatiotemporal structures can hardly be operated without involving individuals (e.g., spacetime points, events, and objects), this does not imply that these individuals should be taken as ontologically primitive. As explained above, individuals play a *heuristic* role allowing for the description of geometric structures, which is not indicative of any ontological priority. It therefore seems that the main difference between the *symmetric* and *asymmetric* A-theories of time is not ontological, but structural (or geometric).

One immediate consequence of the revisited picture is that GBT is no longer to be seen as an ill-conceived hybrid. As a reminder, GBT is traditionally depicted as an intermediate between the polar opposites of eternalism and presentism, with the consequence that this theory accumulates the flaws that are identified in the two traditional models (cf. Sider, 2001: 12, Miller, 2013: 347). But, since eternalism and presentism are no longer to be defined in terms of whether the future and the past exist (because nothing in actual intuition answers to this question), GBT is not a *mere* mixture of eternalism and presentism (which by itself does not guarantee that GBT escapes the objections that are usually formulated against eternalism and presentism). In the revisited picture, what is essential to GBT is to be *asymmetric*, i.e. to say that necessarily sometimes the structure it describes is not reflection invariant. GBT is thus distinct *by nature* from eternalism and presentism. Another consequence is that eternalism and presentism are no longer to be seen as polar opposites, but rather as similar theories: to the question 'Is there a geometric asymmetry between the future and the past?', they both possibly never answer 'yes'. Surprisingly, this leads to a conclusion close to that of the skeptics, who deny that there is a genuine dispute between eternalists and presentists (cf. Crisp, 2004; Ludlow, 2004; Meyer, 2005, 2011; Miller, 2013; Sider, 2006a). But, whereas the skeptics generally justify their claim by invoking the two interpretations of the universal quantifier 'everything' (tenseless or tensed), the suggestion here is to say that eternalism and presentism are similar with respect to some relevant geometric features.

In the McTaggartian picture revisited, the answer ('possibly never', or 'necessarily sometimes') provided to the question 'Is there a geometric asymmetry between the future and the past?' should therefore be considered an essential component of the classical A-theories of time. However, this is obviously not sufficient. First, as Maudlin (2007: 109–110) makes clear, space could contain some sort of asymmetry, but that alone would not justify, for instance, the claim that 'Space is open'. The openness of the future underwrites claims such as 'Anything can happen' or 'History is not written beforehand', while a generic spatial asymmetry would not underwrite such locutions. Second, although the geometric component provides a clear distinction between *symmetric* and *asymmetric* theories, it does not allow us to distinguish between the various forms these theories may adopt. For example, it does not allow distinguishing between eternalism and presentism (which are both *symmetric* theories), nor between GBT and SBT (which are both *asymmetric* theories). Therefore, in order to get a complete categorization of the various A-theoretic views, another component, which does not concern the geometry but the evolution of the model, should be considered essential: the answer ('yes' or 'no') that A-theories provide to the question 'Is temporal becoming (i.e. the creation of new things in the present) real?'.[32] This second component allows one to distinguish between two new groups of A-theories of time: the pure becoming-views (e.g., presentism, GBT)[33] and the non-generative-views (e.g., SBT and eternalism).

Although the question 'Is temporal becoming real?' is of an ontological nature, it can be pre-theoretically evaluated, because, unlike the question 'Do the future and the past exist?', it concerns existence *in the present* that we therefore experience – by contrast, existence in the future and the past cannot be experienced. Thus, not only do we experience a difference between the future and the past, but we also experience that what is occurring (e.g., moments of times, events, etc.) has just been created – this is the idea that *Temporal Becoming* is intended to unpack. According to Henri Bergson, for example, when we are intuiting time we are *primarily* intuiting novelty and creation: "[…] we understand, we feel, that reality is a perpetual growth, a creation without end. […] Every human work in which there is invention, every voluntary act in which there is freedom, every movement of an organism that manifests spontaneity, brings something new into the world" (1944: 261). Formally, *Temporal Becoming* can be expressed as follows: $\mathbf{S}(\exists x\ \mathrm{H}\neg\exists y\ y = x)$, where H is the operator 'always in the past'; this formula literally means: 'sometimes, some things have never existed in the past'. This seems justified by the fact that we do experience things as moving from non-existence to existence and, although this is compatible with the exotic phenomenon of intermittent existence, the most plausible assumption is that those things have never existed in the past. It is worth noting that 'to create' is a non-transitive verb (as opposed to other verbs, such as 'to avoid' or 'to prevent', which are properly transitive in the sense that they link two noun-phrases to signify some relation between real objects). In that sense, a creation is not an action upon what is created, since what is created (e.g., an event) is not there until the productive process is finished (cf. Geach, 1973: 209). That explains *why* *Temporal Becoming*, which implies creation, is not available to eternalism and SBT: if always, everything has always existed ($\mathbf{A}(\neg\exists x\ \mathrm{P}\neg\exists y\ y = x)$), where P is the 'past-tense operator' (read as 'it was the case that'), then nothing can be created. I thus suggest that the following A-theories should be identified with the conjunction of the two logically independent[34] answers they provide to both of these new questions – one concerning the geometry, the other the evolution of the temporal structure – as follows:

[32] In comparison with, for instance, W.L. Craig's conception (2001: 44), the present conception of temporal becoming is restrictive: it only implies the creation of new things in the present, but not the annihilation of some other things. Moreover, following Broad (1923), Reichenbach ([1956] 1971), Prior (1970) and many others, temporal becoming is conceived as a *mind-independent* phenomenon; this will be justified in the next section.

[33] This label comes from Curtis & Robson (2016: 67).

[34] The fact that the geometric and the dynamic components are logically independent allows, for instance, for the possibility of 'frozen-block presentism', according to which the present is not changing (cf. Price, 2011: 279). However, there is an ongoing debate about whether such 'frozen' A-theories of time are coherent (cf. Correia & Rosenkranz, 2020b).

Eternalism: possibly always the temporal structure of the world is reflection invariant (symmetric theory) & there is nothing such as temporal becoming (non-generative-view)

Presentism: possibly always the temporal structure of the world is reflection invariant (symmetric theory) & new things are created in the present (pure becoming-view)

SBT: necessarily sometimes the temporal structure of the world is not reflection invariant (asymmetric theory) & there is nothing such as temporal becoming (non-generative-view)

GBT: necessarily sometimes the temporal structure of the world is not reflection invariant (asymmetric theory) & new things are created in the present (pure becoming-view).

Classifying the above theories this way offers a number of advantages. First, it provides an illuminating and comprehensive categorization of the A-theories of time. Second, whereas it makes presentism, SBT and GBT inconsistent with the classical B-theory *by definition* (geometric asymmetries and temporal becoming are both *incompatible* with the B-theory), it allows for two kinds of eternalism: *A-theoretic* (the moving spotlight) and *B-theoretic* (B-eternalism). Third, it draws on our intuitions by putting the focus on two questions whose specific answers can be confirmed (or refuted) through non-stipulative methods. Finally, and perhaps most importantly, it reveals the metaphysical singularity of GBT: this theory is not *essentially* a hybrid between two polar opposite views (eternalism and presentism),[35] but is an alternative to two symmetric models that mainly differ with respect to whether they accept (or not) *Temporal Becoming*. Thus, there are good reasons to abandon the traditional way of introducing the A-theories of time in favor of the new claims formulated above. Specifically, whereas the traditional definitions rely on a question – 'Do the future and the past exist? – that calls for speculative answers, the new claims rely on two questions that can straightforwardly be evaluated by means of intuitions: 'Is there a geometric asymmetry between the future and the past?' and 'Is temporal becoming real?'. A-theories should therefore be seen as the combinations of two components: one *geometric*, the other *dynamic* (cf. Fig. 3.4).

	Geometric future-past asymmetry	**Temporal becoming**
Eternalism	Possibly never	No
Presentism	Possibly never	Yes
SBT	Necessarily sometimes	No
GBT	Necessarily sometimes	Yes

Fig. 3.4 Two definitional components for the A-theories

[35] To be sure, the claim that 'GBT is an ontological hybrid between eternalism and presentism' is not false, but it grasps nothing that is true *in virtue of the nature* of GBT (which is primarily an asymmetric theory that accepts *Temporal Becoming*), cf. Ingram & Tallant, 2018, §1.

The situation may nonetheless turn out to be more complex than this for at least two reasons. First, one might argue that the two features above – relevant geometric time-asymmetry and *Temporal Becoming* – do not single out GBT. For example, these two features are compatible with the view that 'sometimes, some things go out of existence', which betrays C. D. Broad's original claim that "[t]here is no such thing as ceasing to exist; what has become exists henceforth for ever" (1923: 69). A suggestion would therefore be to regard a further component as essential to GBT: the rejection of *Annihilation*. Formally, this can be expressed as follows: $\mathbf{A}(\neg\exists x\ \mathrm{F}\neg\exists y\ y = x)$, where F is the operator 'sometimes in the future'. This formula literally means that always, everything will always exist in the future and, therefore, prevents the block's erosion. This suggestion is appealing, provided that one wants to strictly respect Broad's intentions. But, one might want to adopt a more liberal stance and allow for other sorts of theories, e.g., 'Partial-GBT' (as opposed to Full-GBT), according to which *Temporal Becoming* holds whereas sometimes, some things will never exist in the future: $\mathbf{S}(\exists x\ \mathrm{H}\neg\exists y\ y = x)\ \&\ \mathbf{S}(\exists x\ \mathrm{G}\neg\exists y\ \mathrm{y} = x)$, where G is the operator 'always in the future'. It indeed seems that such a theory may, under certain conditions (which are not guaranteed by the two principles above), deserve the label 'GBT', since it may also depict a growth in ontology. Intuitively, for all times t, if there are *more* things created than things annihilated up to t, then we get a growing block model at t. Formally, considering the finite set $C(t)$ of created things up to t, and the finite set $A(t)$ of annihilated things up to t, it seems that the partial theory deserves the label 'GBT' iff, for all t, the cardinality of $C(t)$ is *greater* than the cardinality of $A(t)$. Conversely, if (and only if) the cardinality of $C(t)$ is *less* than the cardinality of $A(t)$, then the partial theory seems to deserve the label 'SBT' (cf. Figure 3.5).

Admittedly, some philosophers might regard Partial-GBT and Partial-SBT with suspicion, insofar as (i) nothing guarantees that the set $C(t)$ is finite, and (ii) the expressions 'growing' and 'shrinking' have a quite different meaning in full and partial theories, so that the labels 'GBT' and 'SBT' assigned to the latter might seem usurped. After all, if $C(t)$ is finite, then given that time is dense, the times are not among the things that are created as time goes by. For comparison, suppose one focuses on the spatial size of things and claims that the view that the size of things constantly grows as time goes by is a version of GBT, it clearly seems that one's claim would be wrong. In reply, although one should acknowledge that Partial-GBT

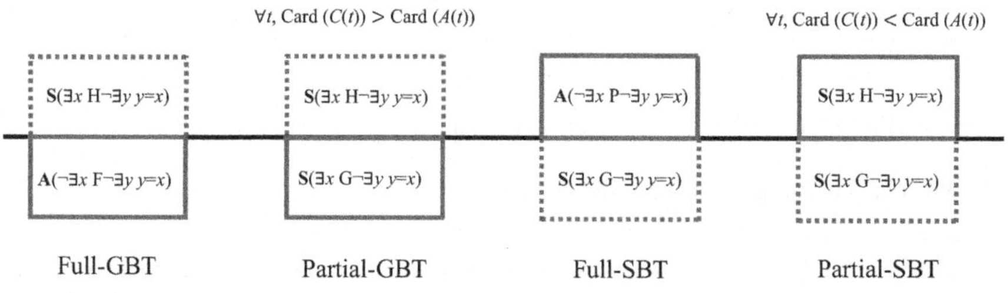

Fig. 3.5 Full and Partial asymmetric structures

goes against many versions of GBT, such as Correia and Rosenkranz's version (cf. Sect. 4.4), it does not go against all versions of GBT. For instance, a growing blocker who says that growth concerns only the occupants of time (e.g., events, people, objects, etc.) would have no *a priori* reason to dismiss Partial-GBT. Of course, one could then question the motivation behind Partial-GBT, since GBT originally rests on the inexorable increase in the ontology of times to explain, for example, the passage of time. However, our question is not whether this view is motivated, but whether it is consistent.[36]

Concerning the symmetric theories, some of them also seem to allow for at least two interpretations, one full and the other partial. In this respect, 'Partial-eternalism' corresponds to the rejection of both *Temporal Becoming* and *Annihilation*: $\mathbf{A}(\neg\exists x\ \mathrm{H}\neg\exists y\ y = x)$ & $\mathbf{A}(\neg\exists x\ \mathrm{G}\neg\exists y\ y = x)$. This form of eternalism is called 'partial', because it is compatible with intermittent existence: e.g., a thing exists for 1 s, then ceases to exist for 1 s, then comes back into existence for another second, and so on from all eternity. The full form of eternalism corresponds to the conjunction of the two following principles: $\mathbf{A}(\neg\exists x\ \mathrm{P}\neg\exists y\ y = x)$ & $\mathbf{A}(\neg\exists x\ \mathrm{F}\neg\exists y\ y = x)$. It is worth noting that Full-eternalism entails *permanentism*, i.e. the view that 'always, everything always exists': $\mathbf{A}\ \forall x\ \mathbf{A}\exists y\ y = x$ (cf. Williamson, 2013: 4). By contrast, presentism intuitively comes in only one flavor: $\mathbf{S}(\exists x\ \mathrm{H}\neg\exists y\ y = x)$ & $\mathbf{S}(\exists x\ \mathrm{G}\neg\exists y\ y = x)$. For example, a presentist will typically claim that every living human being never existed before his birth and will never exist after his death. To sum up, if one thinks of C. D. Broad's version of GBT, Full-GBT, as the only palatable version of the theory, then one has to count the rejection of *Annihilation* among the essential components of GBT (with the difficulty of making it pre-theoretically evaluable). But, if one wants to be more liberal and allow for partial theories, especially Partial-GBT, then the rejection of *Annihilation* must not be regarded as essential component (cf. Fig. 3.6).

A second reason *why* the situation may turn out to be more complex than what the revisited McTaggartian picture suggests is that it seems possible to reconcile an eternalist view with a time-asymmetry. For instance, Maudlin's view of time is explicitly eternalist – "I believe that the past is real [...]. I similarly believe that

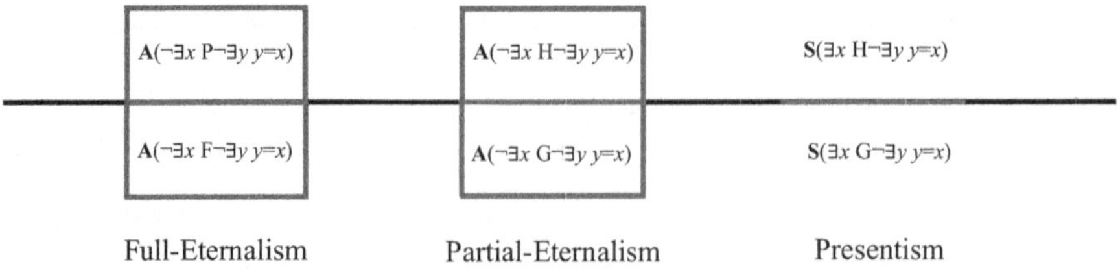

Fig. 3.6 Full and Partial symmetric structures

[36] Nonetheless, Partial-GBT may, for example, be useful when accounting for Maimonides's eschatology according to which some men (those of perfected intellect), but by no means all, will never cease to exist (cf. *Treatise on Resurrection*).

there is (i.e. will be) a single future" (2007: 108–109) – while it admits an irreducible intrinsic time-asymmetry – "I believe that it is a fundamental, irreducible fact about the spatiotemporal structure of the world that time passes" (2007: 107). This suggests that there is room for a midway position between the reductionist accounts of the asymmetry (e.g., those of Boltzmann, 2003, and Reichenbach [1956] 1971, which involve the increase of entropy) and the above 'geometric' account. Huw Price also points in this direction when he confesses that: "[l]ike Maudlin, I am a fan of Earman's Heresy [...]. I think that Earman is right to reject reductionism [...]; but wrong to the extent that he believes that the answer might lie somewhere else" (2011: 286). To that claim, the most plausible reply is that, although Maudlin argues for an intrinsic and irreducible orientation of the universe, his view should not count among the *asymmetric* theories, at least as defined above. After all, Maudlin (2007: 108) explicitly denies that the time-orientation, as he conceives it, has anything to do with the geometric structure of spacetime (though it is not clear, at least to me, what intrinsic structures of spacetime, according to his account, actually yield such orientation).[37]

In a nutshell, this section proposed to reconceptualize the classical A-theories of time in terms that make ineliminable reference to the geometric structures in which individuals (e.g., spacetime points, events, and objects) participate, by using theoretical claims that can subsequently be confirmed (or refuted) by experience. These claims allowed one to distinguish between the *symmetric* and the *asymmetric* A-theories of time. Then, the various forms these theories may adopt were identified thanks to two principles – *Temporal Becoming* and *Annihilation* – in charge of the evolution of the models. The revisited McTaggartian picture showed many advantages, including the power of generating 'partial' versions of eternalism, GBT, and SBT. Another advantage is that the revisited picture changed our perception of GBT, which is no longer seen as a hybrid between eternalism and presentism, but as a theory in its own right. Keeping this in mind, we can now focus on GBT and show that this theory seems suited to successfully accommodate the intuitive asymmetry between the open future and the fixed past (at least as characterized in the previous chapter).

3.5 The Growing Block Theory (GBT)

GBT was first set out in C. D. Broad's *Scientific Thought* (1923). Originally, this theory was characterized as the combination of at least two thoughts: one concerns the geometry and the other the dynamics of the temporal structure of the world. Let's introduce these two thoughts with Broad's own words. First, "[i]t will be observed that such a theory [GBT] as this accepts the reality of the present and the

[37] Some other philosophers share the same concern about Maudlin's account; see for instance Price (2011: 281–283) and Bartels & Wohlfarth (2014: 490–491).

past, but holds that the future is simply nothing at all. [...] The past is thus as real as the present. On the other hand, the essence of a present event is, not that it precedes future events, but that there is quite literally *nothing* to which it has the relation of precedence" (1923: 66). Second, [...] when an event becomes, it *comes into existence*; and it was not anything at all until it had become. [...] The relation between existence and becoming [...] is very intimate. Whatever is has become,[38] and the sum total of existence is augmented by becoming" (1923: 68–69). Although these two thoughts are expressed in ontological terms, they can naturally be read as answers to the 2 definitional questions introduced in the previous section: 'Is there a geometric asymmetry between the future and the past?' and 'Is temporal becoming real?'

To the first question 'Is there a geometric asymmetry between the future and the past?', the growing blocker necessarily sometimes answers 'yes'. She clearly thinks of the time-asymmetry, not as a *natural process* (to be explained in psychological or thermodynamic terms), but as reflecting the geometry of spacetime itself. In that sense, the growing blocker takes the difference between the future and the past to be embedded in the structure of spacetime: just as in the 'forward-branching time' theory (cf. Sect. 2.8), the present is a point-like boundary, around which no reflection symmetry can be operated on fundamental spatiotemporal relations (and derivative spacetime points) without transforming the background structure. But whereas by 'symmetrizing' the forward-branching structure one obtains the backward-branching structure, by 'symmetrizing' the growing block structure one obtains the shrinking block structure. However, since there could have been a first moment of time, a more cautious characterization of GBT is to say that necessarily sometimes the structure it describes is not reflection invariant (asymmetric theory). Thus, according to GBT, what lies below 'the present' axis (viz. spatiotemporally related points) necessarily sometimes fails to be a structural reflection of what is (or precisely is not) beyond this axis. This geometric characterization of GBT does not collapse into an ontological characterization, because of the structuralist approach to the A-theories of time: 'fundamentality' ought to be construed in terms of geometric structures (especially spatiotemporal relations), not in terms of spacetime points (which only play a heuristic role in the description of the spatiotemporal structures). Therefore, assuming that A-theories primarily describe geometric constructions, the essential properties of which are theoretically posited (cf. Sect. 3.4), GBT appears different *by nature* from the traditional models, eternalism and presentism, which both describe structures that are possibly always invariant after a reflection symmetry is operated around 'the present' axis.

To the second question 'Is temporal becoming real?', the growing blocker also answers 'yes'. She clearly thinks of some things (e.g., moments of times, events) as being created in the present. Just as presentism, GBT is therefore depicted as a

[38] As Correia and Rosenkranz point out, the statement "[w]hatever is has become" is too strong, since "[...] there may after all be residents of time that have always existed in the past, or that, although they haven't existed in the past, always in the past, sometimes in the past already existed" (2018: 36–37). There is no *a priori* reason why GBT should exclude such things.

pure-becoming view, which admits that sometimes, some things have never existed in the past: $\mathbf{S}(\exists x\, \mathrm{H}\neg\exists y\; y = x)$. As a result, GBT takes the present to be both *changing* and *distinctive*; it is defined as reality's most recent accretion.[39] By being present, we (as living beings) are therefore located at the leading edge of the block, i.e. at the point of flux itself, where we undergo experiences of the creation of new moments of time.[40] But this is not the end of the story. According to Broad's original version of GBT (Full-GBT), when new things come into existence they always remain in existence: "[t]here is no such a thing as *ceasing* to exist; what has become exists henceforth for ever" (Broad, 1923: 69). Full-GBT therefore requires an additional component to prevent the block's *erosion*, the rejection of *Annihilation*: $\mathbf{A}(\neg\exists x\, \mathrm{F}\neg\exists y\; y = x)$. Intuitively, if one removes this latter condition, there might still be a sense in which the theory deserves the label 'GBT', or more accurately 'Partial-GBT', provided that the sum of what has been created up to a time t is *greater* than the sum of what has been annihilated up to t. This can be expressed by the following formula: $\forall\, t$, Card $(C(t)) <$ Card $(A(t))$.

Taking C. D. Broad's two thoughts seriously brings many benefits. For example, as Broad (1923: 65–66) himself makes clear, they jointly allow for two different senses in which an entity can be said to change its relational properties. Consider the two following cases: (i) Ted, the son of Marie, becomes taller than his mother, (ii) Ted ceases to be the youngest son of Marie. Intuitively, these two cases are very different by nature. In the first case, there are two (partially overlapping) life histories (T and M, say); the earlier sections of T have the relation of 'shorter than' to the contemporary sections of M, while the later sections of T have the relation of 'taller than' to the contemporary sections of M. So, this first case merely reflects a difference of relation between corresponding sections of two existing and temporally-extended entities. In the second case, the change is more substantial. When we state that 'Ted is the youngest son of Marie', we mean that 'There is no entity in the universe such that it is both a son of Marie and it is younger than Ted'. So, when we state that 'Ted has *ceased* to be the youngest son of Mary', we mean that the universe now contains an entity, which did not formerly exist (and therefore could stand in *no* relation whatsoever to Ted), that satisfies both of these conditions. This second case does not merely reflect an evolution in a particular relation between two existing entities, but the coming-into-existence of an entity, a baby, which consequently starts to stand in certain relations to Ted. Whereas GBT has no difficulty in accounting for these two kinds of change, the same cannot be said for symmetric theories (e.g., eternalism, presentism), which possibly always treats the future and the past on a geometric par, and for non-generative-views (e.g., eternalism, SBT), which admits no 'coming-into-existence'.

---

[39] This does not exclude that what is present may lose some intrinsic properties by becoming past (cf. Sect. 3.6).

[40] However, as will be shown (Sect. 3.6), there might be reasons to doubt that we are located in the objective present, i.e. at the leading edge of the growing block.

Moreover, Broad's two thoughts jointly provide an intuitively appealing explanation for (at least some) local asymmetries, such as the increase in entropy.[41] As has been argued, the many attempts to explain (at least some) intuitive past-future time differences thanks to the Second Law of thermodynamics are doomed to fail, because no asymmetry is to be found in the time-reversal invariant fundamental laws of physics (Sect. 2.6). An idea might therefore be to claim that the grounds for the increase in entropy are not to be found in the laws of physics, but rather in the temporal structure of the world as depicted by GBT. Accordingly, the role of the *explanandum* and the *explanans* are reversed: the thermodynamics asymmetry, instead of explaining the asymmetric nature of time, is explained by it.[42] This explanation turns out to be indispensable if one wants to be able to distinguish between the initial and the final state of the universe. As Tim Maudlin makes clear: "[t]he atypical final state is accounted for as the product of an evolution from a generically characterized initial state [...]" (2007: 133). In other words, the increase of entropy away from a low-entropy initial state is the product of a *one-way* evolution; absent such an evolution, there is no way of determining which state is which. This one-way evolution requires that there be a principle, such as *Temporal Becoming*, that provides a direction of time, and can therefore underwrite locutions such as 'The universe evolves towards ever higher entropy'. To put it another way, the evolution of the universe (as we experience it) needs more than phenomenal laws to be explained; it needs principles by which these laws can be interpreted, and *Temporal Becoming* seems well suited to play this role. This explanatory reversal, if successful, blocks the physical possibility of a GBT world in which, although the block grows and the current slice causally brings into existence the new slice, entropy is *decreasing*, and we therefore have no record of past times but apparent records of future times (which turn out to be records, when the future time comes).[43] This is good news, since such a weird world could have served as a counterexample to the claim that the spatiotemporal structure, as described by GBT, is the ultimate *explanans* for our time-asymmetric intuitions. After all, some might have argued that, in such a world, it is the past that would have seemed open and the future fixed.

Finally, and more importantly for our purposes, Broad's two thoughts jointly provide a solid basis to accommodate the asymmetry between the 'open future' and

[41] Another example is famously put forward by Michael Tooley (1997: 111): causation. Assuming that events e_1 and e_2 are causally related, and that this relation requires that e_2 is not real as of the time of e_1, this is best underlain by a pure-becoming view that allows e_1 to exist at a time when e_2 does not exist: GBT.

[42] Admittedly, a detailed account of this explanation remains to be developed. But, following Maudlin (2007), it clearly seems that well-accepted thermodynamical assumptions, such as the microscopic typicality of the initial state (which leads it to higher entropy), require a temporal evolution, while such an evolution is (through *Temporal Becoming*) constitutive of GBT.

[43] Nonetheless, it is not clear how we could have apparent records of future times, if those times never existed. Of course, one might reply that a presentist must account for the fact that we have records of the past, whereas she claims that the past does not exist. But there is a major difference here: the past existed according to the presentist, whereas the future never existed according to the growing blocker.

the 'fixed past', at least understood as a type of worldly (un)settledness to be expressed in terms of there being facts of the matter about what happened but not about what will happen (cf. Sect. 2.9). Indeed, in addition to the failure of physical determinism (cf. Sect. 2.5), the openness of the future requires a structural difference between the future and the past: necessarily, the future should sometimes not be a geometric reflection of the past. The reason is roughly that if the future was always a geometric reflection of the past and, therefore, was also composed of spatiotemporally related points, then no gap in the ontology would be found: facts about what will happen would *supervene* on the spatiotemporal structure, just as facts about what happened *supervene* on it. After all, assuming that nothing is more basic than structure (structuralism), then everything else (including facts about what did and will happen) has a derivative status, so that a structure and its geometric reflection – which share all their geometric properties – cannot allow for different types of facts to supervene.[44] To put it another way, if what lies above 'the present' axis (viz. spatiotemporally related points) always is the geometric reflection of what lies below it, then either both facts about what will happen and facts about what happened supervene, or none of these facts supervene – other possibilities are ruled out by the covariant nature of the supervenience relation. As a result, if one wants to exclude facts about what will happen from our ontology – as is required by the 'no fact of the matter' account of openness – then one has to adopt an *asymmetric* A-theory of time.

An immediate reply might be that the former condition is too strong, since nothing *a priori* prevents presentism – which has been introduced as a *symmetric* A-theory of time – from accommodating the 'no fact of the matter' account of the openness of the future. After all, just as GBT, presentism implies that there are no facts of the matter about what will happen. Therefore, assuming that physical determinism is false, presentism can also allow for a great variety of senses in which the future can be called 'open', including the radical sense according to which time could come to an end (with no ontological commitment to future things standing in the way). In this regard, GBT therefore appears to be in no better position than presentism. This is a fair point. However, only GBT can avail itself of the 'no fact of the matter' account of openness, while keeping the past fixed: for presentism, if the future is open (partly) because there are no facts of the matter about what will happen, so must be the past. In other words, although the 'no fact of the matter' account of the openness of the future is available to presentists, they cannot endorse it without acknowledging that the past is open in the very same sense, in which case they can no longer account for the asymmetry between the 'open future' and the 'fixed past' (cf. Correia & Rosenkranz, 2018: 116). Therefore, insofar as the asymmetry has to be preserved, GBT is much better positioned than presentism to accommodate the open future.

---

[44] As a reminder, supervenience is a covariant relation: "[s]upervenient properties covary with their subvenient, or base, properties. In particular, indiscernibility in respect of the base properties entails indiscernibility in respect of the supervenient properties" (Kim, 1990: 9).

Is it sufficient to adopt an *asymmetric* A-theory of time to accommodate the asymmetry between the 'open future' and the 'fixed past' (as characterized above)? The answer is obviously negative (and so even in an *indeterministic* context). For example, SBT is an *asymmetric* theory, but since it takes all of what exists to have always existed ($\mathbf{A}(\neg\exists x\ \mathrm{P}\neg\exists y\ y = x)$), it cannot be reconciled with an open future (cf. Sect. 2.9). As a reminder, if my grandson (who is neither located in the present nor in the past) exists, then his existence fully settles that I will have children in the future (no matter whether the world is indeterministic, or not), at least assuming that Kripkean necessity of origin holds (cf. Sect. 2.5). Likewise, forward-branching time is an *asymmetric* theory, while this theory was shown incapable of capturing a *genuine* notion of openness that does not arise from our non-neutral perspective on time (cf. Sect. 2.8). Thus, claiming that necessarily the future is sometimes not a geometric reflection of the past is *not sufficient* to accommodate the intuitive asymmetry in openness between the future and the past; what is further required is a principle that involves the creation of new things in the present and, thereby, excludes that all of what exists has always existed, namely *Temporal Becoming*: $\mathbf{S}(\exists x\ \mathrm{H}\neg\exists y\ y = x)$. Specifically, *Temporal Becoming*, when combined with physical indeterminism, allows that some new things are created in the present, while (at least) some of them are not made inevitable by how things located in the present or the past of now are or were (and therefore excludes that all of what exists was *predetermined*). In a nutshell, assuming physical indeterminism, the *combination* of some specific geometric properties *plus Temporal Becoming* makes GBT better positioned than its rivals to accommodate the asymmetry between the 'open future' and the 'fixed past'.

From a wider perspective, it thus seems that GBT does a good job of vindicating the way the temporal structure of the world is pre-theoretically apprehended. Specifically, it satisfies our implicit representation of the future's being open. First, GBT depicts temporal reality as asymmetric, which provides a ground for our intuition that the future differs from the past in various respects, including openness and fixity. Second, GBT is the only theory that can avail itself of the 'no fact of the matter' account of openness, while keeping the past fixed. For, if presentists say that the reason why the future is open is that there is no fact of the matter about what will happen, they must acknowledge (on pain of arbitrariness) that the past is open in the very same sense; after all, they think that there is no fact of the matter about what happened either. So, assuming that the openness of the future is a case of worldly unsettledness that must be expressed in the terms of the 'no fact of the matter' account (cf. Sect. 2.9), it appears that GBT has a decisive advantage over presentism: it preserves the asymmetry that grounds our intuition. Of course, this is not enough to explain the emergence of certain of our practices, emotions, attitudes, and observations that are time asymmetric. This would require further psychological investigations. But this is enough to give a metaphysical basis to the intuition (understood as a pre-theoretic representation of the temporal structure of the world) that is manifested by these practices, emotions, attitudes, and observations.

An observation could be that our phenomenology and practices can be explained by the fact that we do not know what the future will be like. For instance, the reason why it makes good sense to deliberate about what to do in the future but not in the past, or to think that we can causally influence future events (insofar as they counterfactually depend on us) in a way that we cannot influence past events, is that we have past records and no future records. This observation is reinforced by the fact that ontology does not seem to be relevant in that matter: we have no evidence that the future does not exist; so, this ontological feature can hardly be involved in the explanation of our phenomenology and practices. A difficulty might then be the following: since this kind of epistemic explanation is just as available to eternalists and presentists as it is to growing blockers, it may seem that GBT has no advantage over the main competing theories (viz. eternalism and presentism), at least in this respect. As a reply, I do not deny that knowledge asymmetry may play an important role in explaining our phenomenology and practices, but I insist that this kind of explanation is at best *intermediate*. It has then to be explained *where* the knowledge asymmetry comes from. In this last respect, GBT has an immediate explanation that is neither available to eternalists nor to presentists: the knowledge asymmetry is grounded in the geometric features of the spatiotemporal structure. One might reply that there may be other explanations. But, since the laws of nature are time-reversal invariant, and there is nothing fundamentally time-asymmetric in these laws, it is hard to see what explanation eternalists and presentists could resort to. Perhaps they could invoke the increase of entropy, but, once again, this at best postpones the problem: given that there is no directedness in fundamental physics, where does the thermodynamic asymmetry in time come from? (cf. Sects. 2.3, 2.6).

However, despite these attractive features, GBT (at least as defined above) is subject to (at least) four sorts of objections, which one may group as logical, theological, epistemic, and scientific. First, logical objections purport to show that there is something incoherent about GBT *per se*; for example, for the block to literally 'grow', some say, there must be a sensible answer to the question 'At what speed does the block grow?', while such an answer can hardly be found. Second, theological objections highlight the apparent tension between GBT (at least when combined with indeterminism) and God's foreknowledge. Third, epistemic objections contend that GBT leads to radical skepticism; in particular, GBT would provide no basis for knowing that we are living in the present, i.e. on the leading edge of the block. Fourth, scientific objections claim that both the asymmetric nature of GBT and the principle of *Temporal Becoming* are incompatible with our best science, and so would demand a radical revision of the account of temporal structure provided by physics. In order to provide a defense of GBT, these four sorts of objections will be addressed. In particular, logical, theological and 'soft' scientific objections will find concise answers within the remainder of this section; epistemic objections will be the subject of further investigations in Sects. 3.6, 3.7, 3.8, 3.9, and 3.10; finally, the fourth chapter will be entirely devoted to the study of the tension between GBT and Relativity. Let's focus on logical objections.

First of all, the so-called 'rate objection' was first imagined by C. D. Broad (1923) himself, and further developed by some B-theorists of time, such as Smart (1949: 484), Williams (1951a: 463–464) and Price (1996: 13). The rate objection can be formulated as follows: taking seriously the idea that the block grows, it must make sense to ask how fast it grows, which does not seem to be a sensible question.[45] As an analogy, when we ask how fast a car is driving, we ask for a rate of change specified in terms of units of length (for some interval of length taken as a unit) and units of time (for some interval of time taken as a unit). For example, if the car drives at 30 mph to the south (and maintains a constant rate), then after one hour the car will be 30 miles further south. To ask how fast a car is driving is therefore to ask how far the car will have gone when a certain period of time has passed. Of course, if GBT is true, the change that takes place as the block grows is not a spatial movement, and so when we ask how fast the block is growing we are not asking for a rate of change in terms of units of length and units of time. But then what are we asking for? As Huw Price says: "[s]ome people reply that [the block grows] at one second per second, but even if we could live with the lack of other possibilities, this answer misses the more basic aspect of the objection. A rate of seconds per second is not a rate at all in physical terms. It is a dimensionless quantity, rather than a rate of any sort" (1996: 13).

In reply, although one should acknowledge that there is no better answer than saying that the block grows at one second per second, one can argue that this answer is perfectly meaningful and, therefore, that it does not require any objectionable concession. When we ask how fast the block grows, we must mean to ask how the temporal state of things will have changed after a certain period of time has passed. In one hour's time, for example, how will Barack Obama's temporal position have changed? Clearly, Barack Obama will be one hour further into the future, one hour further from his birth, and one hour closer to his death. Admittedly, this answer is not interesting (or informative), but that is only because the question itself is not interesting (cf. Broad, 1959: 766). So, the block does indeed grow at the rate of one hour per hour, one second per second, or 3600 seconds per hour, etc. What is supposed to be objectionable about this answer? Price claims that a rate of change in terms of units of time and units of time is not really a rate, but rather a "dimensionless quantity". But why should one believe this? As an analogy, consider the notion of a fair rate of exchange between currencies, which is usually defined by equality purchasing power: "[…] a fair exchange of euros for dollars is how many euros will purchase exactly what the given amount of dollars will purchase, and similarly for yen and yuan and so on" (Maudlin, 2007: 112). If one asks 'What is a fair rate of exchange of dollars for dollars?', the only sensible answer that can be provided is: one dollar per dollar. This answer is perfectly meaningful – if you think it is not, "[…] imagine your reaction to an offer of exchange at any other rate!" (Maudlin *id*). Thus, just as an exchange rate of one dollar per dollar is *not* free of any specified currency, a rate of one second per second is not a dimensionless quantity.[46] In a similar vein, Bradford Skow (2012: 388–389) argues that the rate at which the period of a pendulum changes can be expressed in seconds per seconds.

[45] The original versions of the argument concerned the more general idea that time passes.

[46] According to Maudlin, Huw Price's confusion rests on the belief that "[…] the units in a rate can 'cancel out', like reducing a fraction to simplest terms" (2007: 113), which is not the case. Maudlin takes the example of π, which is defined as a ratio of a length (of the circumference of a Euclidean circle) to a length (of the diameter). The ratio is length to length: length does not 'cancel out'.

Another logical objection has to do with the point-like boundary conception of the present, which seems to leave no room for non-instantaneous events entirely located in the present.[47] After all, supposing that a non-instantaneous event *e* occurs at *t*, when *t* is present, *e* can only bear one of the 3 following relations to *t*: "[…] either (1) *t* is the last moment of *e*'s existence (i.e. *e* has just finished occurring at *t*), or (2) *t* is the first moment of *e*'s existence (i.e. *e* has begun to occur at *t*), or (3) *e* is presently occurring at *t* by straddling *t* (i.e. *e* has some temporal parts that are earlier than *t*, and some that are later than *t*)" (Diekemper, 2013: 1097). According to Joseph Diekemper, (2) and (3) are no options for GBT, since no event exists (or has temporal parts) later than the present moment. This leads to the counterintuitive conclusion that, given GBT, all non-instantaneous events are past. However, contrary to what Diekemper claims, it is not clear that GBT fails to allow for (2) and (3) – although these two options do indeed require that future (and therefore non-existent) temporal parts compose *e*. At least, provided a distinction (that will be detailed in Sect. 4.7) between two senses of 'exist' – the 'straightforward sense' in which instantaneous temporal parts exist at a time, and the 'derivative sense' in which non-instantaneous entities exist at a time – it seems that, even in a non-eternalist context, events can exist at the present time *t* without having all their parts at *t*. What is merely required is that one of their temporal parts exists at *t* (cf. Lombard, 1999). It therefore seems that speaking of 'present non-instantaneous events' could make sense, even for growing blockers.

Let us now turn to the theological objections. The most famous of them, which has been left pending in the previous chapter (cf. Sect. 2.4), is freely inspired from theological fatalism (i.e. the thesis that God's foreknowledge and human freedom are incompatible): GBT combined with indeterminism is meant to imply an open future, while this seems incompatible with God's foreknowledge. As a reminder, if there is a God who knows the entire future infallibly, then the future is fixed, because nothing can happen that can make God mistaken in his beliefs that the future will unfold in such or such a way. However, just as there are compatibilist strategies to reconcile God's foreknowledge and human freedom, there must be strategies to reconcile God's foreknowledge and the 'open future' view. In this regard, it might be argued that this theological objection rests on a confusion between two different notions: to be true, and to be inevitably true.[48] In brief, if the statement ⌜There will be a sea-battle tomorrow⌝ is true, God knows it; if it is false, God knows it. But knowing that the statement ⌜There will be a sea-battle tomorrow⌝ is true, say, does not make the statement inevitably true and, therefore, does not undermine the 'open future' view. Before detailing this strategy, it is worth saying a bit more about the relation between God and GBT.

---

[47] Consider the spatial analogy of a border between spatial regions: "[t]he border is not a spatial region itself, it does not contain objects; rather, it carves up the spatial regions which themselves contain objects. Similarly, the present is the temporal border between past and future" (cf. Diekemper, 2013: 1100).

[48] This solution was suggested to me by Richard Glauser.

First, let us agree that if God exists, he probably exists outside of time, since Christian tradition has always held that time had a beginning (Augustine, 2006: book 11, chap. 13). It therefore seems more appropriate to speak of God's omniscience rather than of God's foreknowledge. Second, one may wonder whether God's timeless (or atemporal) existence is compatible with GBT. The answer seems to be 'yes'. As Peter Geach, for instance, puts it: "[i]f God sees the world as it is, and the world is temporal and changing, then God must see the world as temporal and changing" (1973: 213). It can therefore be argued that God witnesses the growth of the four-dimensional block, and hence that the future does not exist, even from God's perspective. Third, to the question 'How God's omniscience is to be understood in this context?', a natural answer is the following: God knows the truth-value of *all* statements that have a truth-value, including the truth-value of *all* future contingents that have a truth-value. As a reminder, it was argued that future contingents should not be regarded as exceptions to Bivalence (they *now* have a classical truth-value), while this represents no threat to the 'open future' view (cf. Sect. 2.9). Accordingly, God knows whether the statement ⌜The number of stars is even⌝ is true (or false) and, more interestingly, he also knows whether the future-tensed statement ⌜There will be a sea-battle tomorrow⌝ (the truth-value of which is not *predetermined* by the present or the past) is true (or false).

Yet, the latter claim might seem at odds with the 'open future' view: suppose that God knows that the statement ⌜There will be a sea-battle tomorrow⌝ is true, since knowledge is factive, it seems that no peaceful alternative can take place tomorrow and, therefore, that the future is fixed (at least with respect to sea-battles). But is it really the case? Arguably, a clarification of what the factivity of knowledge amounts to allows one to answer 'no' to the latter question. Let me explain. Whereas the present truth-value of future contingents is not epistemically accessible to us, the same cannot be said for God. Assuming that future contingents are bivalent, the attribute of omniscience allows him to know which of them are true and which are false. So, admittedly God knows that the statement ⌜There will be a sea-battle tomorrow⌝ is true (supposing that it is). However, the factivity of knowledge states that 'X knows that *p*' entails '*p*', not that 'X knows that *p*' entails 'inevitably, *p*'. Therefore, although God knows that the statement ⌜There will be a sea-battle tomorrow⌝ is true, his knowing does not make this statement inevitably true and, therefore, does not undermine the 'open future' view. Otherwise, the openness of the future would rest on the mere epistemic fact that *we* (as human beings) ignore the truth-value of future contingents, which was shown unacceptable (cf. Sect. 2.3). This seems corroborated by the obvious fact that many contingent truths can be the subject of knowledge. For instance, I know that the statement ⌜There are two beers in the fridge⌝ is true; but my knowing that this statement is true does not make this statement inevitably true. There could have been 3 beers in the fridge, after all. Thus, once one has stated that the factivity of knowledge entails true and not inevitable true, it seems that the 'open future' view, which is implied by GBT *plus* indeterminism, and God's omniscience can be retained altogether.

Let's now focus on the scientific objections against GBT. The most striking ones concern the unreality of the future, which seems to require an objective notion of absolute simultaneity forbidden by the theories of Relativity (cf. Putnam, 1967; Rietdijk, 1966). However, the tension between GBT and Relativity deserves special attention and some additional material to be properly understood. It will therefore be treated at length in the last chapter of this book. Other scientific objections concern the principles at work in the definition of GBT, especially *Temporal Becoming*. It might, for instance, be argued that *Temporal Becoming* cannot serve as a definitional principle of a metaphysical theory that aims to describe the temporal structure of the world, since this principle was shown to be *mind-dependent*. According to Adolf Grünbaum (1967), for example, *Temporal Becoming* requires a notion of presentness that is not some physical attribute of events, but depends on mind-possessing organisms experiencing events at a time *t* such that at *t*, these organisms are *conceptually aware* of experiencing these events at *t*. Yet, as Grünbaum claims, this notion of presentness is 'scientifically untutored'; one proof is that two events – such as a stellar explosion that occurred several million years before *t*, and a lightning flash originating only a fraction of a second before *t* – may both be qualified as occurring 'now' by an observer. Therefore, contrary to what the definition of GBT requires, *Temporal Becoming* cannot be thought as reflecting a process that occurs *objectively*, i.e. irrespectively of its relation to conscious organisms, since no event can be called 'present' without being experienced.

However, one may doubt that these scientific facts support Grünbaum's analysis of presentness in terms of a conceptualized awareness of one's own experiences. It indeed seems that these facts can be interpreted in a way that is perfectly consistent with the view that presentness is a *mind-independent* feature of events. As, for instance, Quentin Smith puts it: "[t]hese facts can be taken as suggesting that commonsense ascriptions of presentness to physical events are often in error about *when* these events are present" (1985: 111). In that sense, perhaps the conscious organisms in question are just mistakenly ascribing the time at which the visual effects of the stellar explosion are present to the explosion itself. If so, then the only revision that such scientific facts require is that "[...] the presentness ascribed by commonsense to the stellar events should be taken as a presentness of their visual effects" (Smith *id*).[49] In other words, if conscious organisms are mistaken about associating 'absolute simultaneity' with presentness, then the relevant revision need not be that presentness be regarded as *mind-dependent*, but that the ascriptions of presentness to a physical event *e* be construed *relativistically*, so that '*e* is now' means '*e* is now in this reference frame'. After all, "[i]f the simultaneity of two events is relative to a reference frame and yet mind-independent [as Grünbaum believes], then there is no reason to doubt that the now cannot be also" (Smith *id*).

A related, though more general, scientific objection is to claim that, since sciences (and especially contemporary physics) do not take into account *Temporal Becoming* in their explanations, the explanatory success of GBT – which takes

[49] This can be calculated in accordance with the following formula: "[...] the time at which the visual effects are present minus the distance in light-years to the stellar events" (Smith, 1985: 111). For example, if I am now experiencing visually a stellar explosion 10 light-years away, it can be inferred that this stellar event was present 10 years ago.

Temporal Becoming to be a definitional principle – is threatened. After all, as Hans Reichenbach puts it: "[i]f there is Becoming, the physicist must know it" ([1956] 1971: 16). This objection looks like the one raised against the *fundamentality* of the asymmetry between the 'open future' and the 'fixed past' (cf. Sect. 2.3). As a reminder, the latter objection was framed as follows: since neither the spatiotemporal model favored by physicists (the 'block universe' view), nor the fundamental laws of physics seem to reflect any asymmetry in time, it must be concluded that the asymmetry is at best a non-fundamental phenomenon, at worst an illusion. In the next chapter, I will provide some reasons to believe that *Temporal Becoming* could be restored in the physical context by a dynamics (the so-called 'classical sequential growth' dynamics), which usually enriches the causal set approach to quantum gravity (CST). But, for the moment, I can only repeat what has been said in the previous chapter: there is no reason (neither analytic nor empirical) to think that *all* fundamental phenomena are taken cognizance of by physics. Philosophers like Yuval Dolev (2006) and Mauro Dorato (2008), for instance, have warned against the so-called 'exclusivity dogma', namely the view that "[…] if something is not part of the ontology of physics, then it is not part of the world" (Dolev, 2006: 189).

At this point of the discussion, GBT seems to provide a friendly environment to accommodate various intuitive past-future time asymmetries, including the asymmetry between the 'open future' and the 'fixed past'. The reason is roughly that GBT (i) has itself an asymmetric structure, in the sense that necessarily sometimes it does not treat the future as a geometric counterpart of the past (asymmetric theory), and (ii) allows for the contingent creation of new events in the present (pure becoming-view). Moreover, GBT appears to have the resources to overcome some immediate objections stemming from logic, theology, and contemporary sciences. However, the hardest part remains to be done. First, some philosophers have notoriously argued that GBT leads to an absolute form of skepticism, according to which we (as thinking subjects) have no way of knowing *where* we are temporally located. Second, GBT has often been accused of contradicting relativistic physics, by requiring, for example, an objective notion of absolute simultaneity. These two objections will be examined in that order. Specifically, in the following sections, I will argue that the former objection rests on an uncharitable rendition of GBT. This will give me the opportunity to explain *how* the existence in the past should be conceived, and *why* it sharply differs from the existence in the present. Then, in the next chapter, I will show that, even if GBT turns out to be inexpressible in a relativistic framework (which is doubtful), one can question the credentials of the theories of relativity and argue that our most fundamental physics is rather to be found in the nascent theories of quantum gravity, while some of them seem compatible with GBT.

3.6 A Skeptical Challenge for GBT

GBT is often criticized for not being a *viable* alternative to presentism because of the so-called 'epistemic objection'. This objection was first pressed by Craig Bourne (2002) and further developed by David Braddon-Mitchell (2004) and Trenton Merricks (2006). In its original form, this objection purports to show that GBT leads to absolute skepticism about *where* we are temporally located. In particular, GBT would provide no basis for saying that our time is the objective present. But it is also possible to sharpen the epistemic objection in order to obtain an even more problematic conclusion: GBT would imply that we are, almost certainly, located in the objective past. In other words, the growing block's edge is most likely located in the future of now. Although there are distinctive ways in which the epistemic objection can be formulated, we will focus on the following one.

GBT (at least in its full form) implies, through the rejection of *Annihilation*, that everything that is either past or present exists. For example, Napoleon exists (although he is not located at the present time), as well as everything that concerns Napoleon, such as his beliefs. Among Napoleon's beliefs is presumably the belief that he is located in the present when, for example, he is crowning the Emperor[50]. It indeed seems that, just as we believe that we are located in the present at the current moment, Napoleon believes that he is located in the present when he becomes the first Emperor of the French. In other words, GBT seems to imply not only that thinking subjects located in the objective past exist, but also that they believe that the time they exist at is the objective present. Yet, we obviously know that these thinking subjects are wrong in holding this belief, since we succeed them. In particular, we have no doubt that Napoleon is located in the past (he died two hundred years ago, after all). Of course, that does not mean that Napoleon has never been right to believe that he was located in the present: his belief was true in 1804, but is clearly false in 2022.

Given this, the question that Craig Bourne (2002), David Braddon-Mitchell (2004) and Trenton Merricks (2006) have asked to growing blockers is: 'Assuming that GBT is true, what guarantees that *we*, as thinking subjects, are located in the present?'. Perhaps, in the year 2120, some people look at us just as we look at Napoleon, and think: 'these naive people believe that they are located in the present, while they are embedded in the past!'. In other words, we are in no better epistemic position than thinking subjects located in the objective past who are wrongly believing that they are located in the objective present, since "[…] we would have all the same beliefs […] even if we were past" (Bourne, 2002: 362). GBT seems therefore to lead to skepticism about our temporal localization. Specifically, this theory seems to provide no reason to believe that we are located in the present time.

Worse, the probability that our time is the objective present (and therefore that we are right to believe that it is) is vanishingly small. After all, the objective present might well be located tomorrow, next year, or five billion years beyond the current

[50] It should be remembered that on December 2, 1804 Napoleon crowned himself.

moment, so that we are actually located in the objective past. Since all these alternatives should be regarded as equally likely (our beliefs would in each case be the same, after all), the hypothesis that our time is the objective present is almost certainly false (cf. Braddon-Mitchell, 2004: 200). In other words, the only possibility that we are located in the present does not carry any weight (in terms of probability) in the face of the multitude of possibilities that we are actually located in the past, so that we can assert without a doubt that we are living in the past. These two implications – skepticism about our temporal localization, and the quasi-certainty of being localized in the past – look absurd and must, according to some, lead us to reject GBT.

There are several ways to deal with this objection. First, it may be noted that there is something paradoxical in arguing that GBT leads to absolute skepticism about *where* we are temporally located and, at the same time, that GBT implies that we are (almost certainly) located in the objective past. It seems that these two implications cannot both be true: either a theory is guilty of generating doubts, or it is guilty of generating counterintuitive certitudes. But, clearly, if GBT implies that we are (almost certainly) located in the objective past, then it is simply wrong to claim that this theory provides no basis for knowing *where* we are temporally located: we are in the past, there is (almost) no doubt about this! Of course, one could reply that, although GBT implies that we are (almost certainly) located in the objective past, this does not rule out any form of skepticism; proponents of the epistemic objection seem mainly concerned with our *exact* location. For example, it might be argued that GBT provides no basis for saying whether we are located in the recent or the distant past. But, even if one concedes this, it is still wrong to claim that we have no way of knowing that our time is (or, in this case, is not) the objective present. Obviously, this preliminary remark does not put an end to the debate since, even taken on an individual basis, these two implications remain problematic for GBT. But it indicates that the epistemic objection might rest on some sophistic premises that may (and will) be challenged.

Second, it may be noted that the epistemic objection does not merely concern GBT, but is equally applicable to every A-theory of time that distinguishes between the notions of *existing at the present time* and just *existing*.[51] For example, the epistemic objection is equally applicable to the moving spotlight theorists. In a certain sense, the problem is even more serious for them, since not only could their theory imply an infinite number of possibilities that we are located in the objective past, but also an infinite number of possibilities that we are located in the objective future. In an A-eternalist context, the probability of being localized in the objective present is therefore even lower than in GBT. By contrast, presentism seems to be immune to the epistemic objection: if no other times exist, then there is no puzzle in knowing that we are currently in the objective present (cf. Bourne, 2006: 24,

[51] B-theories of time are not affected by the epistemic objection, due to the indexical nature of 'present' on these accounts (cf. Cameron, 2015: 8).

Braddon-Mitchell, 2004: 199, Heathwood, 2005: 50, Zimmerman, 2008: 216).[52] Of course, pointing out the flaws of other theories does not put GBT in a better position; but it shows at least that if we take the epistemic objection seriously, then we are compelled either to accept presentism or to abandon the idea of an objective present, which sounds suspicious. It would indeed be quite surprising that such a skeptical challenge, though embarrassing, could undermine any attempt to defend a non-presentist A-theory of time, even though sometimes some facts are indeed surprising.

Third, one could regard the epistemic objection as completely harmless. After all, as its name suggests, this objection is of an epistemic nature and, therefore, cannot pose a threat to GBT, which is a metaphysical theory. In that sense, GBT may perhaps lead to absolute skepticism about *where* we are temporally located (or even to the quasi-certainty that we all are temporally located in the past), but this does not imply that GBT is false. What we can (or cannot) know about our temporal location has no impact on the temporal structure of the world. Assuming that the world is such as growing blockers claim, we might have no choice but to accept skepticism as an unfortunate consequence. This reply sounds acceptable, but since my defense of GBT rests on a concern for accommodating a basic intuition we have regarding the nature of time (the future is open while the past is fixed), it would surely be problematic to end up with a theory that implies the truth of positions as counterintuitive as skepticism. Worse than this: if the skeptical conclusion is accepted, the openness intuition for accepting GBT in the first place is undermined, because the future is open only when one is at the present – if I am deep into the past of the block, the future (or at least my immediate future for as long as the present is in my future) is closed.

Finally, and perhaps more interestingly, one may want to properly address the skeptical challenge raised by the epistemic objection. To this end, many options have been considered. I will develop and discuss three of the most interesting ones: Merricks (2006), Forrest (2004), and Correia and Rosenkranz (2018). Then, I will submit my own solution based on the continued existence of bare particulars. From a more general perspective, the presentation of my own solution to the skeptical challenge raised by the epistemic objection will give me the opportunity to explain (i) *how* existence in the past should be conceived, and (ii) *why* existing in the past differs sharply from existing in the present.

---

[52] Nonetheless, Cameron (2015: 23) argues that the epistemic objection faces *every* A-theory of time, not only the non-presentist ones. According to him, appealing to the fact that presentism entails that we are now in the objective present does not help to solve the epistemic problem of *how* we know that we are in the objective present (unless we can come to know that presentism is true, which seems just as difficult as knowing that we are now in the objective present). Cameron's solution to the epistemic objection, which is available to every A-theorist of time, is to adopt an externalist account of knowledge according to which we do not have to have any subjective evidence that the time we exist at is the objective present in order to know that it is so (cf. Cameron, 2015: 49–50). Of course, the main weakness of Cameron's solution is that it crucially depends upon whether externalism is a viable theory of knowledge, while many epistemologists think it is not (cf. Pappas, 2014).

3.7 Three Unsatisfactory Attempts to Meet the Skeptical Challenge

A first solution to the skeptical challenge is due to Trenton Merricks (2006), who argues that the epistemic objection relies on an uncharitable interpretation of what beliefs, such as ⌜I am sitting here at the present time⌝, are about. After all, assuming that a belief about the present cannot occur instantaneously, the growing blocker might have no choice but to concede that such beliefs are never (not even for an instant) true, which sounds implausible.[53] Merricks therefore proposes to distinguish the '*objective* present' (i.e. the growing edge of reality) from the '*subjective* present' (i.e. an indexical, like 'here' or 'this place'). According to him, growing blockers should say that Napoleon's beliefs like ⌜I am crowning the Emperor at the present time⌝ are always about the *subjective* present, so that such beliefs can be true even though Napoleon is not on the edge of reality. Similarly, growing blockers should say that everyone else's beliefs about 'the present' are in fact beliefs about the *subjective* present, which prevents GBT from implying that every belief about the present is almost certainly false. In brief, Merricks argues that GBT should be regarded as a hybrid between the A- and the B-theory of time: growing blockers should agree with B-theorists that 'the present' is typically an indexical (i.e. the present is typically *a matter of perspective*), and they should agree with A-theorists that, in addition, there is an *objective* present (the growing edge of reality) used when GBT itself is being discussed.

However, as Merricks himself points out, the distinction between the '*objective* present' and the '*subjective* present' detracts GBT from its original purpose: to provide a natural view of time. The distinction between the '*objective* present' and the '*subjective* present' calls for other distinctions, especially between the '*objective* future' (i.e. the time which is not yet part of being) and the '*subjective* future' (i.e. the future that follows the *subjective* present).[54] In particular, growing blockers must acknowledge that, just as our typical beliefs about the present are in fact beliefs about the *subjective* present, our typical beliefs about the future are in fact beliefs about the *subjective* future. For example, we are certainly right to believe that ⌜The discovery of a cure for cancer is in the future⌝, provided that by 'the future' we mean 'the *subjective* future'. After all, for all we know, this discovery is in the *objective* past! As a result, it is only philosophers of time, while they are discussing GBT, who use 'the future' to mean 'the time beyond which nothing exists'. But this is clearly problematic, since it is *our* ordinary beliefs about the future (not philosophers'

---

[53] It might be objected that beliefs are dispositions and therefore that they do not 'occur'. However, this objection rests on a confusion between 'dispositional beliefs' and 'occurrent beliefs': "[t]he occurrent belief comes and goes, depending on whether circumstances elicit it; the dispositional belief endures" (Schwitzgebel, 2019: §2.1).

[54] Growing blockers must also distinguish the '*objective* past' (i.e. the past that precedes the *objective* present) from the '*subjective* past' (i.e. the past that precedes the *subjective* present) (cf. Merricks, 2006: 106).

ones) that GBT is meant to guarantee by stating that nothing exists beyond the present! It therefore seems that as soon as GBT distinguishes between two notions of the present – objective and subjective – it betrays the ordinary intuitions that initially made this theory attractive.[55]

Another reason *why* Merrick's suggestion is not appealing is that it makes every typical belief about the present *trivially true*, while – assuming that non-present people may also have such beliefs – we clearly think that they are wrong in holding them. Specifically, if all our typical beliefs about the present are in fact about the *subjective* present (which is to be understood as an indexical that merely refers to 'the moment at which *these* beliefs are held or uttered'), then we cannot be wrong in holding these beliefs. People at any location in spacetime are trivially right to believe (supposing that they do) that they are located in the present, since all that this belief requires to be true is to be held or uttered. Yet, supposing that Napoleon has the belief that the time he exists at is the present, it seems that he is wrong. Napoleon is obviously not located in the present and, therefore, he is wrong to believe (supposing that he does) that he is. In other words, not only do we think that we are right when we believe we are located at the present time, but we also think that people existing at other times (e.g., Napoleon) are wrong in holding this belief. It therefore seems that as soon as one admits that we can have the belief that we are located at the present time without being actually located at *that* time, one should allow for the possibility of being wrong in holding this belief. In brief, it seems that either only people being actually present can have the belief that they are located in the present, or every person can have this belief (including non-present ones) but with the risk of being wrong in holding it.

A second solution to the skeptical challenge is due to Peter Forrest (2004), who claims that the epistemic objection relies on a mistaken assumption, namely that both past and present beings are *conscious* and, therefore, presumably think that the time they exist at is the objective present. According to his highly controversial 'dead past hypothesis', consciousness is a phenomenon which emerges only at the edge of the growing block; it is a by-product of what Forrest calls the "causal-frisson" (2004: 359). If we believe that we are located in the present then we are necessarily right to believe it, since, if we were located in the past, we would not believe anything – we would be zombies (devoid of consciousness). As Forrest puts it: "[l]ife and sentience are, I submit, activities not states. Activities only occur on the boundary of reality, while states can be in the past. […] The past is […] dead" (2004: 359). Thus, according to Forrest, Napoleon exists and he is like us in some respects (e.g., he is a physical entity just as we are), but he is not like us in all respects, as, in particular, he is not conscious. Of course, this is not to say that Napoleon has never been conscious; he was conscious when the time he exists at

[55] Of course, one way of replying is to deny that GBT is meant to do justice to our ordinary beliefs. After all, C. D. Broad's original intention was merely to account for the change of an event in a non-contradictory way, not to capture ordinary beliefs (cf. Broad, 1923: 65–66). But, one again, this reply is not available to me, since my partial defense of GBT is based on accounting for the intuitive asymmetry between the 'open future' and the 'fixed past'.

was the boundary of reality, since it is precisely the fact of being on the boundary of reality that gave him consciousness. However, in 2022 Napoleon is a zombie, while we are not. We can therefore be sure (by introspection) that we are located in the present (cf. also Curtis & Robson, 2016: 77–78).

Nonetheless, although Forrest's response certainly overcomes the epistemic objection, the 'dead past hypothesis' brings its own range of problems. First, this hypothesis seems incompatible with GBT's claim that there always was an edge of reality, including at times when there was no consciousness at all (e.g., 4 billion years ago): if consciousness were a by-product of the 'causal frisson', then consciousness should be observed *at all times*, which is obviously not the case. Of course, Forrest might reply that the 'causal frisson' is merely one of the many necessary conditions for consciousness to emerge. But then, it is not clear what further conditions must be met. Second, the 'dead past hypothesis' entails that presentness generates consciousness in a non-trivial sense. There are indeed many philosophers who argue that "[…] there is no interesting connection between consciousness and presentness" (Meyer, 2016: 151). Worse, there are even philosophers who defend the exact opposite view, i.e. that if there were no judgmental awareness, then there would be no presentness: "[a]n event's occurring now depends on someone's being judgmentally aware of it now" (Baker, 2010: 32, see also Grünbaum, 1967: 17).[56] This is certainly not the place to settle this debate, and we do not want GBT to commit us to any controversial theory of the emergence of consciousness; so it seems preferable to reject the 'dead past hypothesis'.

A further objection against the 'dead past hypothesis' is most famously put forward by Christopher Heathwood (2005): if the 'dead past hypothesis' is true, it makes it difficult for growing blockers to explain how statements attributing consciousness to past beings can be true *now*. In particular, given the classical conception of the grounding requirement on tensed truths (the truth-value of tensed truths must be grounded in how things located in the present are), it seems that *nothing* can make true a statement such as ⌜Napoleon *was* conscious when he crowned the Emperor⌝ (which is surely a present truth, though it is expressed in the past tense). After all, according to the 'dead past hypothesis', things are such that Napoleon exists but is not conscious of crowning the Emperor.

However, it seems that Correia and Rosenkranz's relaxed conception of the 'grounding requirement on tensed truths' (introduced in Sect. 2.9) allows one to solve this issue: since the truth-value of a tensed statement does not need to be grounded in how things located in the present are (e.g., a future contingent might well be said to be true *now* in virtue of how, at some future time, things will be), a past-tensed statement might well be said to be true *now* in virtue of how, at some past time, things have been. In particular, it seems that the truth of the statement ⌜Napoleon *was* conscious when he was crowning the Emperor⌝ merely requires that things have once been that way, i.e. that Napoleon has once been conscious of

[56] This does not necessarily imply that the existence of time itself requires consciousness. Baker (2010: 32) argues even without conscious observers, there would still be instants of time that are ordered by the earlier-than relation (cf. also Meyer, 2016, pp. 145–147).

crowning the Emperor, which is something Forrest explicitly endorses. As has already been said, Forrest argues that past beings were conscious when the time they are at was the 'boundary of reality'. It therefore seems that Correia and Rosenkranz's relaxed conception of the grounding requirement on tensed truths provides the resources needed to protect the 'dead past hypothesis' from Heathwood's objection.[57]

Finally, a third solution to the skeptical challenge is due to Correia and Rosenkranz (2018), who argue that the epistemic objection rests on an uncharitable rendition of the growing block view: "[...] to say, on the one hand, that the past is real (exists), and hence that so are (do) the events that once occurred, is not to say, on the other, that past events are still occurring" (2018: 89). There indeed seems that behind each formulation of the epistemic objection lies the presumption that, according to GBT, past beings (e.g., Cesar, Napoleon) still believe that they are in the objective present[58] – which sounds absurd, since these people are long dead! This presumption clearly distorts the tensed metaphysics underpinned by GBT: the block is not like a "[...] multi-storey building, with lower floors corresponding to the more distant past, where what happens on each floor is still happening, even if it is not happening on the last floor" (2018: 89). The solution proposed by Correia and Rosenkranz is to regard 'occurring' as a temporary property that an event, such as 'Napoleon is having conscious thoughts', once had at some time earlier than now, but no longer has now (without this event having ceased to exist). It indeed seems that C. D. Broad's original view can allow for this solution, since it is anyway committed to temporary properties (that an event once had, but no longer has), such as *being new*. Therefore, contrary to what is presumed in the epistemic objection, it does not follow from GBT that if once Napoleon believed himself to be in the objective present, he is still believing so.[59] Assuming that there are temporary properties, it can be argued that, whereas we believe that our time is the objective present, Napoleon (who still exists

[57] Forrest's response to Heathwood's objection is quite different from the one I have outlined: he claims that it is not surprising that two statements, such as ⌜Napoleon was conscious when he crowned the Emperor⌝ and ⌜Napoleon crowned the Emperor⌝, have very different truth-conditions: this is just part of what makes consciousness special (cf. Forrest, 2006).

[58] Correia and Rosenkranz (2018: 86) acknowledge that the presumption that lies behind some formulations of the 'epistemic objection' (e.g., Bourne, 2002: 362) might be less radical: while past beings no longer believe to be in the objective present at that time, their beliefs are nonetheless answerable to how things stand at that time. However, this second presumption also fails: what needs to be taken into account in order to evaluate past beings' beliefs is only how reality was back then. In particular, the fact that reality has since then grown is irrelevant for the assessment of past beings' beliefs (cf. Correia & Rosenkranz, 2018: 94).

[59] Nonetheless, as Correia and Rosenkranz argue, not only does Napoleon still exist, but so does his belief. However, this does not imply that Napoleon's belief to be in the objective present is false *now*: beliefs of tensed propositions do not change their alethic status depending on the changes in worldly conditions (cf. Evans, 1985: 349–350). In this sense, the fact that, back then, reality was such that Napoleon's time was the last time (i.e. the edge of the block) is the only one relevant for the assessment of Napoleon's belief to be in the objective present. Therefore, there is no sense at all in which Napoleon is now mistaken: if Napoleon's belief was knowledge at the time this belief was entertained, it still is a piece of knowledge now (cf. Correia & Rosenkranz, 2018: 92).

in the past) once had this belief but no longer has it. In short: events can cease to occur without ceasing to exist.

However, although one may agree with Correia and Rosenkranz (and with Forrest) that past beings no longer have belief about the present (or about anything else), it is far from clear that one can always distinguish between the existence of an event and its occurrence. As Peter Geach puts it: "[o]bviously we cannot take this seriously: an actor can be distinguished from his appearance on the stage but we cannot distinguish an event on the one hand and the occurrence or emergence or appearance or taking place of the event on the other hand" (1973: 210). Of course, Correia and Rosenkranz might reply that their theory does not involve non-occurring events, but only events that do not occur at all times at which they exist. However, this reply is not satisfying. This becomes evident when considering a particular event, such as a pain (e.g., a headache): how could a pain exist at a time without being painful at that time? This sounds like a category mistake. A pain, at least as defined by the International Association for the Study of Pain, is "[…] an unpleasant sensory and emotional experience associated with actual or potential tissue damage, or described in terms of such damage" (IASP, 1994: 209).[60] In other words, pains are *subjective*; their existence depends on feeling them. There is no time at which pains exist without being felt. Yet, Correia and Rosenkranz explicitly integrate into their ontology pains that are no longer painful. As they put it: "[one might want to] reject the idea that insofar as the past pain still exists, it still is painful, just as we reject the idea that insofar as WWI still exists, people are still dying in the trenches" (2018: 90). Thus, although Correia and Rosenkranz's solution to the skeptical challenge rests on a fair point (GBT does not imply that events that once occurred are still occurring), it must be rejected on pain of generating paradoxical entities (such as pains that are not painful at all times they exist).

3.8 An Anti-Essentialist Picture of Kinds

Generally speaking, it seems absurd to claim that when objects and events pass from being present to being past, they merely change with respect to their A-properties (i.e. they do not undergo any other alteration whatsoever). As Dean Zimmerman puts it: "[*e*]*verybody knows* that when events and things 'recede into the past' they are very different from the way they are when present" (2008: 221). It is therefore *uncharitable* to presuppose – as Zimmerman (2011), Sider (2011), Merricks (2006) and others do – that GBT *implies* that dead people are currently believing things. If it was really a consequence of GBT, there would not be a single philosopher to defend this theory! Hopefully, GBT offers room for arguing that objects and events,

[60] I take the example of 'pains' because it is, in my opinion, the most striking one. But there are good reasons to think that this objection can be extended to any type of event (including non-subjective ones), since an event is generally defined as "[…] anything that happens, takes place, or occurs" (Simons, 2003: 357).

when they are present (i.e. on the edge of the block), are somehow different from the way they are when past. Of course, this difference could be thought as being merely *extrinsic*. For example, it might be argued that present and past things merely differ with respect to their temporal location: for objects and events, to become past is just to cease to be located *at the last time*. However, although this is a straightforward option for growing blockers, it leads to the same category mistake that was found in Correia and Rosenkranz (2018). For, to say that the difference between present and past things (e.g., present and past pains) is merely *extrinsic* is to say that things do not *intrinsically* evolve by becoming past, and therefore that they belong to the same natural kind (e.g., pains) when present and past, while past pains are not painful anymore. Indeed, since membership in a natural kind is (partly) grounded in sharing natural properties, which are usually taken to be *intrinsic* properties (cf. Bird & Tobin, 2017: §1.1), it seems that no two entities sharing *all* their intrinsic properties can belong to different natural kinds. This is not acceptable: either a pain is painful at each time it exists, or it is not a pain. That is why it seems preferable to maintain that becoming past involves alterations in a thing's *intrinsic* properties, to such an extent that it ceases to belong to its natural kind: a former pain is *neither* a pain that no longer exists (presentism), *nor* a pain that still exists but is not painful anymore (Correia and Rosenkranz), it is no longer a pain.

This option was considered by McTaggart himself when he wondered whether "[...] the change consisted in the fact that an event ceased to be an event [...]" (1908: 459). Although McTaggart promptly rejected this option, on the ground that "[a]n event can never cease to be an event" (natural kind essentialism), it is worth reconsidering it. Natural kind essentialism is the view that membership of a natural kind is essential to its members: if *a* belongs to kind *K*, then it is an essential property of *a* that it belongs to *K* (cf. Fine, 1994; Kripke, 1980). According to this view, a pain is essentially a pain, and gold is essentially gold. Hence, a pain is always if anything a pain, and gold is always if anything gold.[61] However, natural kind essentialism (at least as stated above) is presumably too strong. For example, in chemistry, it is readily acknowledged that "[...] a nucleus of neptunium-239 may undergo beta decay, in which one of its neutrons emits an electron leaving a proton" (Bird & Tobin, 2017: §1.3). As a result, the nucleus in question has one more proton and, therefore, is no longer a nucleus of neptunium-239 but a nucleus of plutonium. Yet, it is intuitively the *same* nucleus (in the numerical sense of the term) that persisted through this transformation. The nucleus has thus retained its identity while undergoing a change of natural kind.[62]

---

[61] It must be acknowledged that the move from '*x* is essentially *F*' to '*x* is *F* whenever it exists' is substantial. Proponents of natural kind essentialism might therefore be more inclined to move from '*x* is essentially *F*' to '*x* is necessarily *F* if it exists'.

[62] As Bird & Tobin (2017: §1.3) have made clear, this example is consistent with the thesis that it is essential to neptunium that its nuclei have 93 protons whereas it is essential to plutonium that its nuclei have 94 protons. In other words, the rejection of natural kind essentialism is compatible with the claim that kinds themselves have essential properties.

Another example concerns biology, and especially Ernst Mayr's influential biological species concept. According to this concept, species are "[…] groups of actually or potentially interbreeding natural populations, which are reproductively isolated from other such groups" (1942: 120). As Mohan Matthen (2009) points out, this definition allows for the creation of new species by the advent of reproductive isolation, which will typically split existing populations.[63] Consequently, the existing organisms in at least one (but presumably both) of the newly isolated (sub-) populations will belong to a new species. As Matthen puts it: "[…] species membership is relational, and consequently, an organism can change its species during its lifetime […]" (2009: 95). Assuming that species are natural kinds, this case offers a second example of particulars changing their kinds.[64] Taking the two above counterexamples seriously, it seems that nothing *a priori* precludes that events may continue to exist, when no longer present, *as a different kind of entity*. For instance, 'Napoleon is having conscious thoughts' can be an event when it occurs in the present and be identical to something else when it is located in the past, without anything having ceased to exist. This option has two immediate advantages: (i) it seems compatible with GBT (since it involves no annihilation) and (ii) it does not generate paradoxical entities (such as events that do not occur at certain times they exist).

Moreover, our own experience seems to match with this anti-essentialist picture. We perfectly know *what* some intrinsic properties are like when they are present, especially the properties we grasp through our own conscious experience. For example, we all know *what* a pain, a sound, or a color feels like. By contrast, a past pain, a past sound or a past color are not being felt, and so they lack this phenomenal quality. Of course, as Dainton puts it, "[…] we do not know what the intrinsic character of a past experience is like, simply because as soon as an experience ceases to be present it is no longer experienced. But [he continues] we can be confident that, whatever this intrinsic character is like, it is different from that of present experiences" (2010: 20). Intuitively, the same reasoning can be applied to material things: "[…] although we know quite a lot about their causal and structural properties (e.g., the shape, size and mass of an electron […]), we have no knowledge of their non-structural intrinsic properties (e.g., what the intrinsic nature of a proton is like in itself). Given this ignorance, we certainly cannot rule out the possibility that the non-structural intrinsic properties of material things undergo changes when they become [past]" (Dainton *id*).

Taking seriously the idea that a thing undergoes an *intrinsic* alteration when it becomes past, the question that immediately arises is: 'Which intrinsic property is altered?'. Timothy Williamson (2013) suggests that a thing that becomes past loses

[63] Reproductive isolation is the mechanism according to which "[…] the habits, habitat, physiology, and genetics of organisms that belong to a single species enable them to recombine their genes with those of others of the same species but not with those of any other species" (Matthen, 2009: 96).

[64] Many other counterexamples can be found in Bird and Tobin (2017).

its *concreteness*.[65] In brief, he conceives reality's dynamic nature as grounded in temporal shifts from non-concreteness to concreteness and from concreteness to non-concreteness. According to him, a pain is thus something before and after it occurs (namely a non-concrete pain), and it is concrete only while it is occurring. Of course, this does not imply that a non-concrete pain is an abstract entity (*pace* Sider, 2001: 127). As Williamson makes clear, 'non-concrete' and 'abstract' are not to be treated as synonyms. 'Abstract' has its own positive paradigms, such as numbers and directions. So, when a pain disappears (and therefore becomes past), it is not trans-formed into an abstract object; it merely ceases to be concrete. However, although this solution may look attractive, it has (at least) two drawbacks that I mention now.

First, since Williamson provides no perspicuous account of his notion of concreteness, his proposal is not fully intelligible. As he himself concedes: "[t]he term 'concrete' is used informally throughout this book. For present purposes, we need not decide between various ways of making it precise (being material, being in space, being in time, having causes, having effects, …)" (2013: 6). Without any further specification, it is therefore hard to make sense of Williamson's proposal. Second, assuming that non-concrete things do not occupy any spacetime region (which sounds both conventional and respectful of Williamson's intention), it is not clear that this proposal is available to GBT. Of course, one can conceive a Williamsonian version of GBT, according to which future things do not exist, present things exist *concretely*, and past things exist *non-concretely*. However, unless the past is to be thought as an empty place, this version of GBT renders *superfluous* the existence of the past, at least understood as a physical location (i.e. a place where concrete things can be located). If the present is taken to be the unique temporal location for concrete things, then there is no reason to assume that other temporal locations (such as the past) exist – these temporal locations would always remain empty, after all. It therefore seems that if one adopts the Williamsonian version of GBT, one must reject the view that some things can be temporally located elsewhere than in the present. This makes this version of GBT (i) very similar to certain versions of presentism, such as 'thisness presentism' (according to which past things survived by abstract entities: their thisnesses),[66] and (ii) difficult to reconcile with contemporary physics (according to which concrete things can be located elsewhere than *now*, just as they can be located elsewhere than *here*).

A Williamsonian reply could be that, although Socrates is not in space *now*, i.e. Socrates is not located at the present three-dimensional slice of the four-dimensional manifold (this is presumably the sense in which Socrates is non-concrete *now*), he *does* occupy a spacetime region *now*, in the sense that he *was* in space some years

[65] This does not preclude that concreteness can also be seen as an *extrinsic* property, especially when 'being concrete' is identified with 'being located in spacetime'. However, this is not the kind of view we are interested in here, since, as has been argued above, becoming past must involve alterations in a thing's *intrinsic* properties (to such an extent that it ceases to belong to its natural kind). Therefore, extrinsic concreteness cannot do the job.

[66] Cf. Adams, 1986, Keller, 2004, Ingram, 2016, 2018.

ago. In this regard, the spacetime region that Socrates occupies *now* is, for instance, wholly before the spacetime region that Napoleon occupies *now*. However, although one must acknowledge that 'being in space' (unlike 'being in spacetime') can be introduced as a relative notion and, therefore, that a thing can be located in spacetime *now* without being located in space *now*, this reply misses its target. The reason is that this should not be the sense in which growing blockers say that, for instance, Socrates *is* in spacetime. I mention two arguments in favor of this claim. First, even presentists (who deny the existence Socrates) would agree that Socrates *was* in space some years ago and, therefore, that he is located in spacetime *now* (in the above sense). Yet, since GBT and presentism might disagree on what is located in spacetime *now* (e.g., either both Socrates and Obama, or only Obama), it seems that 'being in spacetime' should have a more specific meaning within GBT, which entails that everything that is in spacetime *now* is also in space *now*. Second, supposing that a thing can be in spacetime *now* without being in space *now* (as permitted by the above sense of 'being in spacetime'), it follows that the hundredth President of the United States (who does not exist according to GBT) does occupy a spacetime region *now*, simply because he *will be* in space in a few years. This is not acceptable. Growing blockers should therefore argue that everything that is located in spacetime *now* is located is space *now*, and hence that Socrates is still (and will always be) a concrete entity (*pace* Williamson).

So, if 'concreteness' is not the right answer to provide to the question 'Which intrinsic property is altered when a thing becomes past?', what could it be? As a first approach, the most plausible answer is 'it depends': although a thing that becomes past undergoes an alteration in some of its intrinsic properties, the intrinsic properties concerned by this alteration depend upon what kind of entity is becoming past.[67] There are indeed good reasons to believe that a person, a pain, and a stone do not change in the same way by becoming past, since they did not share the same bedrock of intrinsic properties in the first place. For example, while a person who dies (and therefore becomes past) presumably loses the property of being conscious, it would be absurd to extend this to all entities, since most of them have never been endowed with consciousness. Likewise, events seem to no longer occur when past, while this cannot be said about people or stones.

For now, all that can be said in order not to fall back into the pitfalls of Correia and Rosenkranz's theory is that whatever the sort of alteration a thing that becomes past undergoes, this implies that this thing no longer belongs to its *natural kind* (rejection of natural kind essentialism). A person that is no longer conscious is no longer a person; an event that no longer occurs is no longer an event; and a pain that is no longer painful is no longer a pain.[68] Yet, just as it is the same nucleus that

[67] There are other views on which things are intrinsically different (when past) then they were when present. One can mention, for instance, Cameron (2015), and Miller (2019).

[68] Here, it is naively assumed that each present thing belongs to a *single* natural kind (because it seems to be the case in the classical Aristotelian view). But there seems to be no *a priori* reason to reject the idea that a thing may belong to several natural kinds. In any case, this would not affect the view that will be developed below.

persists through the addition of a proton to become a nucleus of plutonium, these are the same things that persist through the passage of time to become past. There is therefore nothing such as ceasing to exist; when past, the same things continue to exist, falling under a different natural kind. Given this, the interesting question to ask is not 'Which property is altered when a thing becomes past?' (since the answer varies depending on what kind of thing is considered),[69] but rather 'What remains of a thing that became past?'. For example, 'What remains in 2022 of Napoleon's belief that he is located in the present?'. In order to answer this question, I will introduce, in the next section, the notion of 'bare particular'. This notion will be conceived, in a deliberately open-ended manner, as what is responsible for the continuity of existence of both continuants (people, tables, planets, etc.) and occurrents (events, processes, etc.), through both superficial change (e.g., becoming warm) and radical change (e.g., becoming past). It is worth insisting that bare particulars are merely intended to explain *how* existence in the past should be conceived, not to provide grounds for present truths about the past, since these grounds are already guaranteed by the relaxed conception of the 'grounding requirement on tensed truths' introduced in Sect. 2.9.

Before proceeding any further, it could be objected that involving bare particulars in the debate is not an alternative to Williamson's view, but rather a way of making his proposal definite or precise: the shift from concrete to non-concrete can be understood as the shift from non-bare to bare. Supposing that one can accurately determine what the intrinsic properties are, the loss of which makes a thing turn from non-bare to bare, it seems that nothing prevents Williamson (who is not known as a friend of bare particulars) from claiming that the loss of *these* properties is precisely what makes a thing become non-concrete. In reply, two things can be said. First, it is not clear that this is an objection to bare particularism. As has been said, one of the major drawbacks of Williamson's view is that it is not fully intelligible – it lacks a perspicuous account of the notion of concreteness. So, if bare particularism, while not being an alternative, helps lift the veil of mystery surrounding the term 'concrete', then it is surely worth considering it. Of course, the risk is to end up with two views, only one requiring the pseudo-exotic notion of a bare particular, which seems to be a decisive advantage for Williamson's view. But, as will be shown, the notion of a bare particular is less exotic than it might seem: it merely refers to what resides over and above the properties of any continuant and occurrent whatsoever and, therefore, is an ontological free lunch for those who accept the classical theory of individuation (the substratum theory). Moreover, even if the notions of bareness and non-concreteness turn out to be co-extensive, one might prefer to work with the former, since it echoes a long-standing conception of change, according to which change is possible only if something always persists through any change that occurs.

---

[69] Nonetheless, the question 'What is it for something to become past?' will find an informative answer in the next section: a thing (e.g., a person) that becomes past loses the properties that make it belong to the kind to which it belonged when present (e.g., consciousness).

Second, it is not clear that bare particularism can be understood as a way of making Williamson's view definite or precise. Although it might seem difficult to answer the question 'What are exactly the properties the loss of which makes a thing turn from non-bare to bare?', one thing is certain: it cannot be the properties that make things concrete (at least assuming that only concrete entities occupy spacetime regions). As has been said, GBT requires that (at least) some past things are concrete, otherwise the past would be a superfluous empty place. So, if bare particulars are what remains of things that became past, then they *must be concrete* and, therefore, they cannot be confused with Williamson's past things. Perhaps Williamson could reply that some non-concrete things do occupy spacetime regions. For example, consider the two following things: Wittgenstein's possible daughter and Descartes. According to Williamson, whereas the former occupies no spatiotemporal region whatsoever, there are certain times (e.g., January 1, 1650) at which the latter is spatially located. There therefore is a sense in which Descartes, despite not being concrete, occupies spacetime regions. However, although it must be acknowledged that Williamson's view allows one to distinguish between non-concrete things that have been spatially located (e.g., Descartes) from those that have never been (e.g., Wittgenstein's possible daughter), this example misses its target. The times at which Descartes was spatially located (e.g., January 1, 1650) are the times at which Descartes was *concrete*. So, this example provides no reason to believe that, in Williamson's view, some things are spatially located at the times at which they are non-concrete. In a nutshell, whereas Williamson thinks that Descartes was concrete but is not anymore (Descartes is not concrete *now*), the 'bare particular' theorist thinks that Descartes is concrete *now* and has always been so. There therefore seems to be no extensional overlap between the notions of bareness and non-concreteness.

3.9 Bare Particulars to the Rescue of GBT

If one wants to account for how things located in the past exist, it is worth considering the primitive notion of a 'bare particular'. This notion comes from the substratum theory, according to which "[…] particulars are, in a certain sense, separate from their universals" (Sider 2006b: 387). More specifically, the substratum theory says that particulars and universals are only connected to each other by a relation of instantiation. That means that particulars do not have properties *as parts*; they instantiate them. As Ted Sider puts it: "[t]hey are nothing but a pincushion into which universals may be poked" (*id.*). John Locke speaks of them as the "I know not what" substrata ([1689] 1975, II, xxiii, §2), while Plato (2000) uses the term "receptacles" (*Timaeus* 48c-53c).[70] I personally prefer the expressions 'bare particular', 'substratum', or '*thin* particular' (as opposed to '*thick* particular' which refers to the

[70] For a discussion of whether Plato is committed to bare particularism, see McPherran, 1988.

fusion of a *thin* particular and its universals).[71] In other words, a 'bare (or *thin*) particular' is the mereological difference between a thick particular and its universals. Whereas bare particulars play a predominant role in the individuation debate – they are usually posited to account for the identity and distinctness of particulars (cf. Bergmann 1967, Moreland 1998, Sider 2006b) – they might also, as we will see, be helpful in the philosophy of time.

Importing a notion issued from the individuation debate into the philosophy of time is less crazy than it might seem. After all, in the *Physics* (Book I, chap. 6), Aristotle (1999) himself connects the notion of a 'material substratum' (in which the properties exemplified by a particular inhere) to the question of change. He argues that any change must be analyzed in reference to an invariant substratum. However, it would be a mistake to claim that the notion of a 'bare particular' has an antecedent in Aristotle; for, he explicitly says that everything that exists (with the notable exception of the eternal substance of the unmoved mover) is a compound of matter and form (hylomorphism). Specifically, Aristotle distinguishes two types of change: sometimes a change is just a change in a characteristic or two, as when the cold bronze sphere becomes a warm bronze cube, and sometimes the matter itself changes and we are no longer dealing with bronze at all (cf. Cohen, 1996: 67–68). The first type of change is called 'reciprocal', since there can be a transformation back into the original stuff, whereas the second change is called 'nonreciprocal', since it is definitive. However, both types of change require something that survives the transformation (otherwise, the process would not be called 'a change'), namely a substratum which is responsible for the continuity of existence – this substratum is matter (*hyle*), which is not to be understood as a bare particular, but as a pure potentiality. As Christopher Byrne explains: "[t]hese requirements apply to essential change just as much as to accidental change, for Aristotle insists that there is a persisting substratum in generation and destruction as well" (2018: 47). Of course, it is the second type of change (i.e. 'nonreciprocal change') that will be matter of great concern in the remainder of this chapter, since becoming past is, in every sense of the expression, a definitive process.

Against the substratum theory there is the bundle theory, according to which "[...] particulars are just bundles of universals" (Sider 2006b: 387).[72] Although

[71] The distinction between *thin* and *thick* particulars was popularized by David Armstrong (1997). However, according to Armstrong's specific version of the substratum theory, particulars (which we encounter in ordinary experience) are states of affairs, and *thin* particulars exist only insofar as they can be "abstracted" from the *thick* particulars with which they are associated.

[72] There are at least two ways to be a bundle theorist: "[*u*]*niversalist* bundle theorists take the properties or characteristics that are shared by objectively similar concrete particular objects to be universals. In contrast, *trope-theoretic* bundle theorists hold that that the properties or characteristics that are shared by objectively similar concrete particular objects are themselves particulars, viz., so-called *tropes*, *moments*, or *modes*. Tropes are construed by their proponents as particularized properties or individual qualities, e.g., the particular redness that inheres in a rose. Both types of bundle theorists view concrete particular objects as in some sense composed of, or constituted by, the properties that enter into a particular bundle; only they differ over whether the very same properties can be multiply located in distinct bundles at a single time" (Koslicki, 2018: §1.4).

substratum and bundle theorists agree on much – e.g., they agree that both particulars and universals exist, and that a particular somehow *has* universals – they do not share the same conception of what a particular is. Whereas a bundle theorist affirms that a particular is nothing but all the universals it has (i.e. it is the mereological fusion of all its universals), a substratum theorist denies this. According to the latter, when you take a particular, and you mereologically subtract away all its universals, there is something left: a bare particular. So, assuming the principle of uniqueness of mereological fusion (no universals can have two fusions), the bundle theory and the substratum theory differ sharply over the possibility of exactly similar particulars: whereas a bundle theorist affirms that no two particulars can have exactly the same universals (since a particular is just the sum of its universals),[73] a substratum theorist affirms that distinct particulars *can* have exactly the same universals (since they will have distinct bare particulars, i.e. distinct non-universal 'cores'). The benefits of adopting a substratum theory rather than a bundle theory will be detailed in the next section.

Now, assuming that bare particulars constitute a fundamental ontological category, it might be argued that, although *becoming past* involves alterations in a thing's intrinsic properties (to such an extent that it ceases to belong to its *natural kind*), the bare particular of *that* thing will continue to exist. For example, if the Battle of Waterloo is conceived as a temporally-extended particular instantiating some properties[74] (e.g., having opposed France to the Seventh Coalition, having been won by the Duke of Wellington), then this battle will always be something when no longer occurring: a bare particular. This bare particular is what persisted through the intrinsic alteration that affected the battle of Waterloo when it became past. As an analogy, consider the case of a statue made of bronze, and suppose, for the sake of argument, that bronze is the substratum. The statue is, in that sense, bronze instantiating some properties (e.g., *having the shape of a woman, being cast by Rodin*). Now, suppose that this statue is melted so that all that remains is a warm bronze cube. Clearly, after such a process, bronze has lost some properties (e.g., it no longer has the property of *having the shape of a woman*, nor the property of *being cast by Rodin*). As a result, it can no longer be called 'a statue' (or perhaps only in a non-trivial sense by a bunch of contemporary artists). Nonetheless, bronze survived the transformation. Of course, this analogy has limits: whereas melting is a superficial (or reciprocal) change which does not affect bronze in its proper kind (bronze remains bronze), becoming past is a radical (or non-reciprocal) change which definitely turns every thick particular into a bare particular.

---

[73] The bundle theorist (unless she opts for a trope-theoretic version of her theory) is therefore committed to the controversial 'Principle of Identity of Indiscernibles' (PII), according to which necessarily, any two particulars that have all the same qualitative properties are the same particular.

[74] As far as I understand it, this approximates the classical Davidsonian conception of events (cf. Davidson [1970] 2011). For example, David Papineau has explicitly identified Davidson's events with bare particulars: "[t]here are various possible ways of instituting the required relationship between mental and physical causes. If we think of events as bare particulars, we can say that each particular mental cause is the same event as the relevant physical cause (cf. Davidson's 'Mental Events')" (1990: 66).

To the question 'What remains of a thing that became past?' a plausible answer might therefore be: a bare particular. An immediate objection is that, according to this answer, the proper name 'Socrates' seems to refer to two distinct entities: a thick particular 'T', which has gone out of existence, and a bare particular 'B', which is still with us. This poses at least two issues: (i) it is not clear whether Socrates should be identified with B, T or both B and T, and (ii) it seems that T's going out of existence is incompatible with GBT, the full form of which rejects the principle of *Annihilation*. Let us start with the first issue. At first sight, it seems perfectly acceptable and convenient to say that Socrates *is* T, and that Socrates *is* B. The most natural view is indeed that Socrates came into existence in 469 BC, then lived as a thick particular until he drank the hemlock in 399 BC, and continues to exist as a bare particular since then. To put it another way, just as it seems that it is the *same* bronze that is a statue at t_1 and a warm cube at t_2, it seems that it is the *same* particular (Socrates) that is thick in 350 BC, and bare in 2022 AD. This natural view escapes, by the way, the second issue, since it does not imply any form of annihilation: there is simply a single entity (Socrates) which, although it has existed in two different forms (a *thick* particular and a *bare* particular), has never (and will never) cease to exist.

Unfortunately, things are not so simple, especially because if Socrates *is* both B and T, then it follows that B *is* T, which seems to betray Leibniz's law. To illustrate, suppose that t is a moment at which Socrates was alive: at t, T has the property of *being alive* as a part, but at t, B does not have the property of *being alive* as a part. By the Principle of the Indiscernibility of Identicals (necessarily, if two particulars are identical, then they have all the same qualitative properties), it follows that B is not identical to T. This is analogous to a well-known situation in the metaphysics of constitution, which is usually introduced as 'the puzzle of Tibbles the cat' (cf. Burke, 1996; Geach, 1980; Wiggins, 1968). In brief, suppose that before us stands a cat named 'Tibbles'. Before us is also that part of Tibbles which consists of all of Tibbles except his tail. Let us call that part of Tibbles 'Tib'. Since Tibbles has a tail, but Tib does not, it follows, by the Leibniz Law, that Tib is not identical to Tibbles. Suppose now that Tibbles loses his tail. At this moment, we are inclined to say that Tib is identical to Tibbles. After all, Tib and Tibbles now occupy the same volume at the same time and, as David Wiggins puts it, "[i]t is a truism frequently called in evidence and confidently relied upon in philosophy that two things cannot be in the same place at the same time" (1968: 90). Hence, assuming that identity is not contingent, we are faced with a contradiction: Tib *is* and *is not* identical to Tibbles. Of course, claiming that either Tib or Tibbles has ceased to exist is not an option for growing blockers (at least given the full form of GBT) and, in any case, would be arbitrary. As Michael Burke makes clear: "[t]he identity of a cat surely is not tied to its tail. So Tibbles still exists. But surely Tib has not ceased to exist: Tib lost none of its parts" (1996: 63). So, what should we do?

First, it must be noted that the puzzle of Tibbles the cat does not only concern bare particularism, but any view that analyzes change in reference to an invariant substratum and, thereby, allows things to retain their identity while losing parts. Therefore, if one wants to reject bare particularism on this basis, one should be ready to endorse a view such as mereological essentialism, i.e. "[...] the doctrine that every part of an object, no matter how small, is essential to its identity" (Burke *id*), which brings its own range of problems. For example, how can mereological essentialism account for the fact that common-sense objects persist through change? Considering that bananas ripen and houses deteriorate, how can this view say that they are the same things, if they are not quite the same? Second, there is a battery of well-known solutions to the puzzle of Tibbles the cat, which strongly suggests that the above objection against bare particularism is not definitive. For example, some philosophers opt for four-dimensional entities whose parts may extend in time as well as in space (cf. Heller, 1990; Lewis, 1986; Sider, 2001), some others argue that identity is a contingent relation that may hold at some times but not at others (Gallois, 1990; Gibbard, 1975; Myro, 1985), Peter van Inwagen (1981) claims that there are no such things as arbitrary undetached parts, David Wiggins (1968) distinguishes between the 'is' of constitution and the 'is' of identity, etc. The question is then, 'Which of these solutions are available to bare particularism within GBT?' It would be difficult to provide an exhaustive answer to this question, but it *a priori* seems that at least two solutions could be retained: (a) relativizing numerical identity, and (b) distinguishing between two senses of 'is' (constitution and identity).[75]

However, option (a), which amounts to saying that B was identical to T when Socrates was alive and that B is now distinct from T since Socrates died, has two major drawbacks: (i) it looks *ad hoc* (at least in the present context), and (ii) it forces one to reject the 'constituent thesis', according to which every thick particular has a bare particular (and some properties) as constituents, which is at the core of bare particularism (cf. Bailey, 2012). Of course, this is not to say that these drawbacks cannot be overcome. Niall Connolly (2015), for example, happily rejects the 'constituent thesis'. He argues that the best version of bare particularism takes the relation between a substance (e.g., a tree) and its substratum to be *identity*, and not the relation of constitution. Nonetheless, a better option (which requires fewer concessions) is to preserve the 'constituent thesis', by arguing that when one says that Socrates is T and that Socrates is B (which again is perfectly acceptable and convenient), one actually means that Socrates *constitutes* T and that Socrates *is identical to* B. The distinction between these two senses of 'is' was first highlighted by David Wiggins (1968) and Peter Geach (1980). As, for instance, Wiggins puts it: "[t]he 'is' of material constitution is not the 'is' of identity. The tree is *made of* (or *constituted of* or *consists of*) *W*, but it is not identical with *W*. And '*A* is something over and above *B*' denies '*A* is (wholly composed of) *B*' or '*A* is merely (or merely consists of) *B*.' If *A* is something over and above *B*, then of course $A \neq B$, but the proper point of saying 'over and above' is to make the further denial that *B* fully *exhausts* the matter of *A*" (1968: 91–92). It is however worth noting that, according to Wiggins' constitution view, the substratum *W* which constitutes the substance *T* is not to be under-stood as a bare particular, i.e. as something 'over and above' *T*, but as a 'superinternal relation'[76] that organizes the parts of *T*. So conceived, the relation of constitution is distinguished from identity insofar as it is asymmetric: *W* constitutes *T*, but not vice versa.

[75] The question of the compatibility of GBT and four-dimensionalism will be explored in Sect. 4.7.

[76] The expression comes from Cameron, 2014.

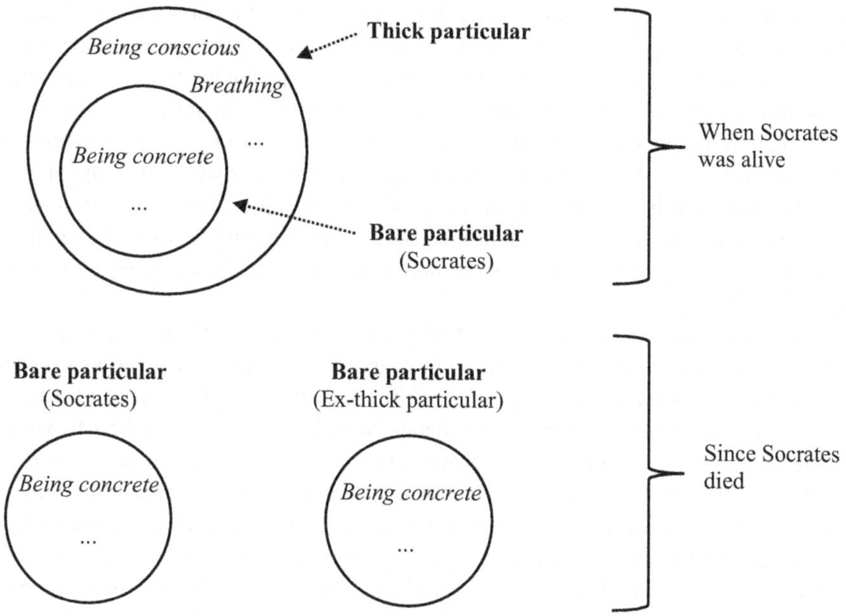

Fig. 3.7 Bare and thick particulars at various times

Now that we have distinguished between the two senses of the word 'is' (constitution and identity), we can return to the case of Socrates and see how this applies to it. As a reminder, GBT (at least in its full form) commits us to the claim that T still exists now, just like B (rejection of *Annihilation*). A question then arises: 'What are the properties that T now has?'. Certainly not those that were specific to Socrates when he was alive (e.g., being *conscious*, *breathing*, etc.). The most plausible answer is that T and B now have exactly the same properties. Taking this seriously, the situation is as follows: B has always been a bare particular (at least since it began to exist); on the other hand, T is bare now but has not always been so: when Socrates was alive, T was a whole constituted by (i) a bare particular (namely B) and (ii) properties such as *breathing* and *being conscious*. The situation is depicted in Fig. 3.7. An immediate objection is that this version of bare particularism is ontologically costly: it generates a colossal number of bare particulars. A reply could be that, from a qualitative point of view, the doctrine remains parsimonious, by keeping down the number of *sorts* of entities (it only involves thick and thin particulars). And, according to Lewis (1973a: 87), it is the only parsimony criterion one should consider.

Another objection is formulated by Andrew Bailey (2012). Roughly, this objection says that if one accepts 'the constituent thesis' (i.e. the thesis that every thick particular has two kinds of constituents: its properties and its bare particular), then one should abandon 'the having thesis' (i.e. the thesis that every thick particular has its properties by having as constituents properties that are instantiated by another of its constituents: its bare particular). Proof is that, taken together, these two theses imply that two co-located entities (the bare particular and its host thick particular) have the same properties, which sounds false. However, although persuasive, this objection exerts no pressure on the above view, since ‘the having thesis’ is not taken to be true. In general, a bare particular does not instantiate the properties had by its host thick particular (rejection of ‘the having thesis’); it merely has these properties in a derivative way. For example, Socrates only had in a derivative way the properties that one usually attributed to him (e.g., *being conscious*, *breathing*, etc.). Does this undermine the view as a version of bare particularism? It does not seem so, since other self-appointed ‘bare particular theorists’ reject ‘the having thesis’ (cf. Wildman, 2015).

The claim that what remains of a thick particular that became past is a bare particular looks attractive for the following four reasons. First, the notion of a bare particular is (at least in itself) *economical*: it is an ontological free lunch for those who accept the classical substratum theory of individuation, according to which particulars are not exhausted by the properties they exemplify, but are associated with a substratum. Second, the notion of a bare particular is *familiar*: it echoes a long-standing conception of change, according to which change is possible only if something always persists through any change that occurs. In this perspective, ‘becoming past’ should just be conceived as a radical (or nonreciprocal) type of change. Third, this view is *compatible* with GBT’s main imperatives: the possibility that thick particulars may continue to exist as a different kind of entity (a bare particular) is not excluded by the rejection of *Annihilation* (i.e. the ontological preservation of what is no longer present), which was characterized as an essential feature of GBT (at least in its full form). Fourth, bare particulars have a great *explanatory power*: they allow for a conception of the past – ‘the bare past’ – in which nothing occurs (e.g., there are no events, no movements, so that no property can be gained or lost) and, therefore, offer an appealing account for the fixity of the past. For all these reasons, it seems that bare particulars should be conceived as what is left of a thick particular when certain of its intrinsic properties were subtracted by the passage of time. Quentin Smith endorses a similar view when he claims that “[past particulars] are ‘bare particulars’ in the sense that they lack nonrelational, monadic properties” (2002: 132). But while Smith’s view (so-called ‘degree presentism’) attributes this ‘bareness’ to the fact that past particulars are only partly real,[77] GBT should not allow for degrees of existence: past and present particulars are just as real.

Unfortunately, as previously said, it seems difficult to identify one unique intrinsic property whose loss would make particulars turn from thick to bare, since thick particulars of various kinds (e.g., events, people, stones) do not share the same bedrock of intrinsic properties in the first place. Nonetheless, it seems possible to provide an informative criterion: the properties that a particular loses when it turns

[77] Specifically, Smith admits degrees of existence: the more an event is temporally distant from the present, the less it is real. The result is that present particulars (or maximal existents) have nonrelational monadic properties and also stand in relations, while past (and future) particulars (i.e. what exists to less than the maximal degree) only stand in relations (cf. Smith, 2002: 132).

from thick to bare are (at least) the ones that make it belong to the natural kind to which it belongs when present. These are the properties in terms of which natural kinds are traditionally defined. For example, in June 1815, if it is the property of *occurring* that makes a thick particular, constituted by the battle of Waterloo, belong to the kind 'events' (Simons, 2003: 357), then it is at least this property that the particular in question loses when it becomes past. Likewise, in the fifth century BC, if it this the property of *being a featherless biped* (*Categories*, 3a 23–5), or the property of *being a rational animal* (*Politics*, 1253a 10), that make a thick particular, constituted by Socrates, belong to the kind 'man', then these are at least these properties that the particular in question loses when it becomes past. In other words, ex-thick particulars are free from all the properties that jointly define the natural kind to which they belonged when present. This is what turns them to bare. This criterion does justice to the idea that both continuants and occurrents continue to *concretely* exist when no longer present (rejection of *Annihilation*), but as a different kind of entity (rejection of natural kind essentialism) – a bare particular – which results from the alteration of some specific intrinsic properties they possess.

Taking this criterion seriously, an immediate question is: 'Do things that cease to belong to their initial kind necessarily belong to a new kind?'. Or, to put it another way: 'Are there particulars free from any natural kind?'. For cases of reciprocal change, the answer is quite obvious. For example, it is clear that a nucleus of neptunium-239 that undergoes beta decay (in which one of its neutrons emits an electron leaving a proton) does belong to a new natural kind after the transformation, namely plutonium. Likewise, given Ernst Mayr's biological species concept, it is clear that reproductive isolation splits existing organisms of the new (sub-)populations into new species, which correspond to new natural kinds. But, the question arises in a more interesting way when asked about things that became past: 'Do bare particulars, such as Socrates or the thick particular he constituted, still belong to a natural kind after Socrates' death? The most plausible answer is 'yes'. Aristotle, for instance, leaves no room for entities that do no not belong to any natural kind. But then, the question is: 'To which natural kind do these particulars belong after Socrates' death?'. The most straightforward answer is 'the kind of bare particulars' (which does not imply that bare particulars are, in themselves, natural kinds). This answer seems partly justified by the fact that Socrates, Napoleon, and WWI have some relevant properties in common, such as *being a particular*, *being concrete*, *having belonged to another natural kind*, etc.[78] Is it the end of the story? Perhaps Socrates and Napoleon should have more in common than Socrates and WWI (though they are all bare particulars). Perhaps there should be a kind to which both Socrates and Napoleon belong, but not WWI. Yet, I strongly reject this suggestion: there is no qualitative distinction among bare particulars. The reason is that all the

[78] Obviously, sharing some properties is not sufficient for constituting a kind (even if it is a necessary condition). But, among the criteria of a natural kind classification that have been listed by Bird & Tobin (2017: §1.1) (to have some properties in common, to permit inductive inferences, to participate in laws of nature, etc.), none of them seem to prevent us from conceiving a kind of bare particulars.

properties that would be required to constitute kinds were lost through temporal change. Of course, this is not to say that one cannot distinguish between Socrates, Napoleon, and WWI. According to the substratum theory, the individuation is ensured by the bare particulars themselves.

Despite their attractive features, Ted Sider observes that there is a complaint against bare particulars: "[they] are widely regarded as the grossest of metaphysical errors" (2006b: 392). For example, it might seem that if a thing has no properties, then there is at least one property that this thing has, namely the property of having no properties, which renders the notion of 'bare particular' incoherent. An immediate reply is that if the objection is that bare particulars have no properties at all, then the objection is just wrong. Once again, the bare particulars involved in the above view of 'how things become past' do not need to be free from all properties, but merely from those that together make *these* things belong to the natural kind to which they belong when present. In that sense, bare particulars do have some properties, such as *being a particular*, *being bare* or *being concrete* – the nature of a bare particular is, by the way, given by the properties it instantiates.[79] In that respect, the expression 'bare particulars' is misleading; it does not mean that some particulars are entirely free from properties, but rather that properties are not constitutive parts of a substratum. As Bradley Rettler and Andrew Bailey make it clear: "[b]are particulars are 'bare' in at least this sense: unlike objects, they have no properties as parts" (2017: §3.2). In other words, the expression 'bare particulars' conveys the idea that the link between a substratum and the properties it bears cannot be a relation of constitution, but must be a relation of instantiation; otherwise the substratum (assuming it has no further constituents) would be just a bundle of properties.

A more charitable interpretation of the objection, then, is that if particulars were wholly distinct from their universals, it would be possible for there to exist *truly* bare particulars (i.e. particulars that instantiate no properties at all), while this proposal is incoherent. Again, *truly* bare particulars would have at least one property, viz. the property of having no properties. In reply, three things can be said. First, one can prevent the possibility of *truly* bare particulars through an appropriate conception of modality (though this does not seem desirable).[80] David Armstrong (1989), for instance, builds the impossibility of *truly* bare particulars into his theory of possibility. Second, a similar objection can be addressed to bundle theorists, since nothing *a priori* excludes the possibility "[…] where no universal is compresent with any universal, not even itself" (Sider 2006b: 392). Third, and perhaps most

[79] An objection might be that only properties that are *parts* of a particular can characterize its nature. But, following Sider (2006b: 390), I do not see *why* it should be so. After all, many properties a bare particular may instantiate are *intrinsic* properties. Moreover, even assuming that there could be *truly* bare particulars (i.e. particulars that instantiate no properties whatsoever), I do not see *what* could prevent us from saying that these particulars have a nature simply by *failing* to instantiate properties. In that sense, all *truly* bare particulars have the same nature, and that nature is "[…] exhausted by the fact that they instantiate no monadic universals" (Sider 2006b: 392).

[80] As will be shown in the next section, *truly* bare particulars might be useful in the metaphysics of spacetime, especially when considering *Ontic Structural Realism*, and in the metaphysics of mathematical entities.

importantly, this objection rests on a confusion between *sparse* and *abundant* properties.[81] In the abundant sense of 'property', each meaningful predicate corresponds to a property; so "[…] if we could predicate 'has no properties' of a thing, then that thing would indeed have a property corresponding to the predicate" (Sider 2006b: 392). But this is clearly not the relevant sense here; rather 'property' is to be understood in the sparse sense: just as *being red* or *being red or round, has no property* does not correspond to a property. As Sider puts it: "[j]ust as a thing can be red or round without having a sparse property of being red or round […], a thing can have no sparse properties without having a property of having no sparse properties. And of course, the substratum theorist's universals are sparse" (*id*).

A further complaint might be that substratum theorists are unable to give a coherent account of the instantiation of universals, which is a major reason for rejecting their theory. As David Lewis puts it: "[c]onsider the predicate 'instantiates' (or 'has'), as in 'particular *a* instantiates universal *F*' or 'this electron has a unit charge'. No one-off analysis applies to this specific predicate" (1983: 353–354). However, although one must acknowledge that most substratum theorists remain silent on the predicate 'instantiates',[82] it is not clear that such an analysis has to be provided. In particular, it seems that substratum theorists can argue that 'instantiates' is part of their ideology – this might even seem required, since postulating a dyadic universal of instantiation (to bind particulars to their universals) would at best postpone the need for primitive predication and, therefore, generate an uneconomical regress. Moreover, it is not clear that bundle theorists are in a better position: they need to take the predicate of 'compresence', i.e. the predicate that relates the universals had by a given particular to one another, as primitive. This indeed seems required in order (i) to say which fusions of universals count as particulars (e.g., there are fusions containing the universals of goldenness and mountainhood as parts, whereas there is no particular such as a golden mountain),[83] and (ii) to prevent that any universal had by a part of a particular is had by that particular (e.g., Geneva is a town, while Switzerland is not a town, though Switzerland has Geneva as a part).[84]

A last complain concerns proper names. Against the descriptivist tradition (cf. Frege 1982), names are typically seen as having no linguistic meaning beyond their reference. Ruth Barcan Marcus (1961), for instance, argues that proper names should be regarded as 'tags', since they refer *directly* to their bearers, i.e. not by way of descriptions. Taking that for granted, the question that arises is: "[…] what it is, if not an associated description, that fixes what a name refers to [?]" (Reimer, 2019: §2.2). The most popular answer to this question is the so-called 'causal theory of

[81] The distinction between *sparse* and *abundant* properties is due to David Lewis (1986: 59–69).

[82] There are notable exceptions: Baxter, 2001, Armstrong, 2004.

[83] The bundle theorist Laurie Paul overcomes this objection simply by denying that composition is unrestricted. According to her, there are no fusions containing the universals of *goldenness* and *mountainhood* as parts (cf. 2002: 579–580).

[84] This objection can be overcome either by denying that the parthood relation is transitive, or by saying that particulars are 'composed' of universals in some sense that does not involve the usual notion of parthood.

reference', according to which (i) a name's referent is fixed by an original act of naming, and (ii) subsequent uses of that name succeed in referring to that referent by being linked to the original act of naming via a causal chain (cf. Kripke, 1980). As Marga Reimer puts it: "[…] speakers thus effectively 'borrow' their reference from speakers earlier in the chain, though borrowers needn't be able to identify any of the lenders they are in fact relying on" (2019: §2.2). This popular answer, however, might seem at odds with bare particularism. Suppose that the proper name 'Socrates' refers to a bare particular. Since bare particulars are typically taken to be causally powerless (causal powers are necessarily connected with natural properties, after all),[85] it might seem that the link between 'Socrates' and its referent is broken. As Keith Campbell puts it: "[a]ll causal action is exerted by way of the properties of things and all effects are effects on the properties of things. The substratum, precisely because it is without properties, including passive powers, ought to be totally immune to all causal activity" (1990: 9). In reply, one should insist on the fact that bare particulars being powerless does not imply that the reference-link is broken.[86] For instance, although Socrates is a bare particular, and therefore is powerless (after having been causally active, through a thick particular he constituted, for more than 70 years), we can still successfully refer to him. The reason is that *our act of referring* is causally linked to the original naming of Socrates (by which 'Socrates' became a rigid designator of that particular). Specifically, Socrates (via his initial baptism) is at the origin of a causal chain (which ensures that later uses of the name 'Socrates' succeed in referring to him), although he is devoid of all the properties in virtue of which he was (in a derivative way) causally active.

At the end of the day, the continued existence of bare particulars seems to offer an elegant story (i.e. a story that does not generate any paradoxical entity) about the kind of change continuants (people, tables, planets, etc.) and occurrents (events, processes, etc.) undergo when they become past. In particular, once an event ceases to be present, it is no longer in any sense occurring and, therefore, it is no longer in any sense an event (rejection of natural kind-essentialism). Of course, this is not to say that something ceases to exist (rejection of *Annihilation*); what remains (and will always remain) from *that* event is a bare particular, i.e. a substratum that is at least freed from the property of *occurring* (supposing that this property is what makes a particular belong to the kind 'events'). Events are thus to be thought as a natural kind to which some particulars do belong when present; by becoming past, the same particulars continue to exist but, since they have undergone an intrinsic alteration (which involves the loss of the property of *occurring*), they are now bare. Is this story sufficient to solve the epistemic objection? The answer is 'not quite'. It cannot be simply because my believing is occurring that I know that this belief is

[85] The view that causal powers are necessarily connected with natural properties (which can be challenged, cf. *Humean contingentism*) is called 'necessitarianism' (cf. Tugby, 2021). The two most popular necessitarianist theories are 'dispositional essentialism' (cf. Bird, 2007) and 'the identity theory' (cf. Heil, 2003).

[86] If it were the case, some might claim that we could not successfully refer to any powerless entity, such as *abstract entities* (e.g., Sherlock Holmes), which sounds suspicious.

located at the present time; I might well ignore that it is occurring, in which case I would not know that it is located at the present time. What is further required to solve the epistemic objection is the claim that, as a matter of general fact, if one's belief is occurring, then one knows it by introspection. Thus, the reason *why* we know that our time is the objective present is that (i) such an event (i.e. 'someone's believing to be present') could not occur in the past (which is to be conceived as a spatiotemporal region exclusively populated by bare particulars), and (ii) we introspectively know that our belief that we are located in the present is occurring.

To be sure, the above conception of the past – 'the bare past' – allows growing blockers to deny that GBT entails that events are still occurring in the past (*pace* Bourne, 2002, Braddon-Mitchell, 2004, Merricks, 2006). Indeed, assuming that bare particulars are what will forever be left of events, GBT coheres with the intuitive idea that an event, such as 'someone's believing to be present', can only occur at the present time. The past is therefore fixed. Accordingly, 'Napoleon's thinking about crowning the Emperor in the present' is *not* occurring in the past; rather, 'Napoleon's thinking' occurred, and what remains of that former event is a bare particular. Then, given that if one's belief is occurring, one introspectively knows it, we can be confident that we are located in the present, i.e. at the leading edge of the growing block, when we think we are (rejection of skepticism). Thus, contrary to what the epistemic objection states, we (as constituents of conscious events) do find ourselves in a far better epistemic position than Napoleon, who is no longer the constituent of any event whatsoever. The situation is summarized in Fig. 3.8.

A quick look at my conception of the past – 'the bare past' – might lead one to believe that my theory is more akin to a version of presentism than to a version of

	Continuants	**Occurrents**
Present	*Thick* particulars being constituted by: (a) A bare particular (b) Various properties: - All the properties essential to the natural kind to which they non-essentially belong (e.g., *being conscious*, if a person; *having chemical formula H_2O*, if water; etc.) - Many other contingent temporary properties.	Temporally-extended *thick* particulars being constituted by: (a) A bare particular (b) Various properties: - All the properties essential to the natural kind to which they non-essentially belong (e.g., *occurring, happening, taking place*, etc.) - Many other contingent temporary properties.
Past	Bare particulars *instantiating* properties such as: - *Being a particular* - *Being an individual* - *Being concrete* - *Having belonged to another natural kind*	Bare particulars *instantiating* properties such as: - *Being a particular* - *Being an individual* - *Being concrete* - *Having belonged to another natural kind*

Fig. 3.8 Continuants and occurrents at present and past times

GBT. For example, it is correct to say that my theory (like Correia and Rosenkranz's one, by the way) leaves no room for events (e.g., the Battle of Waterloo) that are still occurring in the past. It is also correct to say that past beings (e.g., Cesar) are no longer having beliefs. Both of these features would be enthusiastically accepted by a presentist. But one crucial difference between my position and presentism is that I do think the Battle of Waterloo and Cesar still exist concretely in the past. Admittedly, these things exist in a different form than when they were present (since they lost many temporary properties in the process of becoming past), but they remain concrete things (viz. bare particulars), which a presentist would never admit! This offers, by the way, some advantage to my position on presentism, since the conception of the 'bare past', according to which the past exists although nothing happens in it, provides an appealing account for the fixity of the past. As a result, the ontology is clearly increasing according to my theory, as new things (e.g., times, events) come into existence to join the things that already exist, which again seems incompatible with presentism. This conception of the past fits perfectly with the spirit of GBT, which does not imply that events that once occurred are still occurring. Recall Correia and Rosenkranz's metaphor: the growing block is not like a multi-storey building (cf. Sect. 3.7) – to say otherwise would be to misrepresent the tensed metaphysics underpinned by GBT. For these different reasons (and some others mentioned earlier, cf. Sect. 3.5), there seems to be no risk of confusing my version of GBT with presentism.

Finally, one could object that my 'bare particular' view betrays our intuitions, although I suspect that this has more to do with the term 'bare particular' than with the concept itself. After all, this notion comes from the substratum theory, which is arguably the most intuitive theory of individuation (cf. Sect. 3.9); at least, the idea that particulars are, in a certain sense, separate from their universals, seems to be widely shared. Likewise, with respect to the question of change, the idea that for something to change, something (that is responsible for the continuity of existence), thus potentially a bare particular, must survive that change, also seems intuitive (cf. Sect. 3.9). But, let us leave that aside. The important point is that, even if bare particulars should definitely be regarded as exotic entities, this would not detract from the intuitive character of my theory. To be clear, my theory is not intuitive in the sense that it resorts only to intuitive entities – this would be an unreasonable requirement for a metaphysical theory that aims at ruling on the structure of reality. Rather, my theory is intuitive in the sense that it accounts for some basic intuitions we have regarding the nature of time. Specifically, it posits bare particulars for accounting for two intuitive aspects of reality: (i) the fact that existing in the present differs from existing in the past, and (ii) the fact that the past is fixed. In that respect, the label 'intuitive theory' does not seem to be usurped as far as my theory is concerned.

3.10 The Virtues of Bareness

When considering the previous story about *how* things located in the past exist, based on the continued existence of bare particulars, the most important worry to deal with is to justify the claim that there are such entities. After all, many philosophers think that the concept of 'what it is for a particular to be the particular that it is' can be captured by an exhaustive list of all that particular's qualitative properties and, therefore, that ontological parsimony dictates against the postulation of bare particulars. In this section, I will argue that, notwithstanding their bad reputation, bare particulars exist and are required in order to account for how things located in the past exist. As has been said, bare particulars are to be thought as what is responsible for the continuity of existence, through both superficial change (e.g., becoming warm) and radical change (e.g., becoming past). In particular, they are what is left of events (and other kinds of *thick* particulars, such as people or stones) when certain of their intrinsic properties (viz. the properties that make *these* particulars belong to the kinds of entities to which they belong) were subtracted by the passage of time. In that purpose, I will first provide an accurate characterization of what bare particulars are, so they cannot be confused with other entities that can be found in the philosophical literature. Then, I will mention various independent contexts in which bare particulars may be useful. Finally, I will provide some reasons to think that the 'bundle' alternative is false.

The first thing to say is that bare particulars are *individuals*, not properties, and are *concrete* (i.e. they occupy spacetime regions), not abstract *entities*. Consequently, they cannot be confused with *thisnesses*, where the thisness of a thing x is the abstract non-qualitative property of being x (or the abstract non-qualitative property of being identical to x).[87] In this sense, the thisness of a thing x is *not* merely a conjunction of all of x's qualitative properties, but it is x's property of being just *that* thing. For example, the thisness of an event, such as the Battle of Waterloo, is the property of that particular event, and of nothing else. The reason *why* it is worth distinguishing bare particulars from non-qualitative thisnesses is that both of these notions stem from the individuation debate: they are posited as competitive alternatives to the view according to which particulars are just bundles of universals. More specifically, substratum and thisness theorists agree that there may be distinct particulars that do not differ in respect of the universals they possess. But, whereas the former thinks that a particular is individuated by its substratum, the latter thinks that it is individuated by its thisness.

Of course, the nature of thisness is controversial, but one view is that thisnesses are primitive, i.e. they are elements of reality that cannot be reduced to anything more fundamental. Those who believe in primitive thisnesses typically believe that

[87] This is the Scotistic conception of thisness (after Duns Scotus who coined the term 'haecceitas', of which 'thisness' is intended to be a translation). This conception also stands in opposition to the 'bundle' conception of thisness, according to which a thing's thisness is merely the conjunction of all of its qualitative properties (cf. Adams, 1979: 6–7).

a thisness comes into existence with its thing, and continues to exist as long as it is exemplified by *that* thing. But some philosophers, e.g., Robert Adams (1986), Simon Keller (2004), and David Ingram (2016, 2018), go a step further by arguing that the property of thisness continues to exist, even though it is no longer exemplified. As Ingram puts it: "[o]n my view, for any entity x, x's thisness *T* comes into being with x, *T* is uniquely instantiated by x throughout x's existence, and *T* continues to exist uninstantiated when x ceased to exist" (2018: 61). Following these people, it might therefore be argued that non-instantiated thisnesses – rather than bare particulars – are what is left of things (e.g., events) when they become past.

As a result, the 'thisness view' might also seem to provide a solution to the epistemic objection: there are no events, such as 'someone's believing to be present', occurring in the past, since all that remains of past events is their non-qualitative property of thisness. Given that if one's belief is occurring, one introspectively knows it, we (as constituents of conscious events) can be confident that we are at the present time when we think we are. However, this solution does not seem available to GBT (at least in its full form). The most obvious reason is that friends of the thisness view hold that past entities cease to exist while being survived by their thisnesses, which is incompatible with GBT's rejection of *Annihilation*. Moreover, just as with Williamson's proposal, the 'thisness view' renders the existence of the past *superfluous*. As has been said, thisnesses (and therefore non-instantiated thisnesses) are *abstract entities*, while abstract entities are generally taken to occupy no spatio-temporal regions. Thus, unless the past is to be thought as an empty place (i.e. a place where no concrete entities are located), the 'thisness view' can only be accommodated by presentism, which takes the present to be the unique temporal location for concrete entities.[88] It is no coincidence that most philosophers accepting thisness ontology – Adams (1986), Keller (2004), Ingram (2016, 2018) – are presentists.

It should now be clear how bare particulars can be helpful in telling *how* things located in the past exist. However, since bare particulars are often perceived as exotic entities, it might be objected that any story involving them is, without any further justification, inevitably *ad hoc*. In the previous section, one provided some reasons to think that bare particulars are less exotic than it might seem; a further reply might be to show that the 'epistemic objection' is not the only problem bare particulars may help solve. For example, it seems that bare particulars might help growing blockers reply to Kit Fine's McTaggartian argument for the unreality of time (since the rejection of *Neutrality* does not necessarily dispel the contradiction). As a reminder, this argument rests on the claim that, in every A-theoretic ontology (with the exception of presentism), reality is constituted by events with *incompatible* content (at least assuming that reality should allow for its being variegated over time).[89] Specifically, either the totality of events constituting reality are those that obtain at past, present, and future times (MSL), or the totality of events constituting

[88] It is worth noting that the 'thisness view' can also be accommodated by SBT, which takes the present and the future to be the only two temporal locations for concrete entities.

[89] Kit Fine (2005) speaks of "tensed facts" rather than "events", but this does not play a crucial here (provided that we bear in mind that the argument does not commit to an ontology of events).

reality are merely those that obtain at past and present times (GBT), or the totality of events constituting reality are merely those that obtain at present and future times (SBT) (cf. Sect. 3.2). Now, assuming that what remains from past events are merely bare particulars, the argument for the unreality of time is no longer a threat to GBT. The growing blocker can indeed argue that events come into existence in virtue of their substratum being present, but then that it is merely bare particulars (understood as the forever-existing traces of what is no longer present) that exist in the past. Growing blockers are therefore no longer committed to a reality being constituted by events with incompatible content: there merely are events that presently occur and bare particulars that forever belong to the past. There therefore is no time at which two events with incompatible content are both constitutive of reality.

Moreover, bare particulars (especially *truly* bare particulars, i.e. particulars that instantiate no properties whatsoever) seem to be useful in various independent contexts, such as the metaphysics of spacetime. For example, considering the *Ontic Structural Realism* (OSR) of Ladyman and Ross (2007), i.e. the view that the world has an objective modal structure that is ontologically fundamental, it appears that physical relations do not relate particulars with intrinsic qualities, but merely undifferentiated spacetime points. As Ladyman and Ross put it: "[t]here are objects in our metaphysics but they have been purged of their intrinsic natures, identity, and individuality, and they are not metaphysically fundamental" (2007: 130). Although OSR is often regarded as the most plausible alternative to supersubstantivalism (i.e. the view that material objects are identical to spacetime regions),[90] it faces difficulties in distinguishing the relations that hold (i.e. that are instantiated) from those that do not. A solution might therefore be to take the physical relations that are instantiated by *truly* bare particulars. In that sense, spacetime points might be considered as *truly* bare particulars that stand in a fundamental structure of physical relations (cf. Sider 2006b; Schmidt, 2008; Connolly, 2015). This solution provides a reasonable alternative to both antirealism about science (i.e. the view according to which scientific terms are not to be interpreted as referring to anything) and eliminative structuralism (i.e. the view that there are no objects, but merely relations), without threatening the ontological priority of the structure.[91] It will in particular be useful in Sect. 4.6, when we will interpret the causal sets theory (CST) in structuralist terms.

Finally, assuming that *sui generis* mathematical entities exist, such as natural numbers ordered in an arithmetical ω-sequence, a question might be: "[w]hat distinguishes these objects from others, in virtue of which they are numbers?" (Sider 2006b: 393). Of course, many answers can be provided to that question. For example, one might argue that there is a distinctive property – Sider speaks of a "numerical glow" (cf. 2006b: 393) – that is shared by these entities and only by them. However, since it is not clear what such a property would be like (mathematical

[90] Supersubstantivalism is defended by Quine (1981), Skow (2005), and Schaffer (2009).

[91] It should be acknowledged that this solution would probably be rejected by Ladyman and Ross, since bare particulars, as elements of reality, are not known to science (cf. Ladyman & Ross, 2007: 14).

knowledge fails to describe any intrinsic properties of mathematical objects), a better answer might be that it is the relation ordering the ω-sequence that is distinctive (cf. Shapiro, 1997, Sider 2006b, Ladyman, 2005, Leitgeb & Ladyman, 2008). This answer (so-called 'mathematical structuralism'), which can be understood as OSR's counterpart in the metaphysics of mathematical entities, leads us to think of the members of any arithmetical sequence as *truly* bare particulars. Here again, this solution preserves structuralism from both antirealist and eliminativist views, by providing minimal *relata* to fundamental relations.

Now, if the successful uses of bare particulars in various independent contexts is not convincing enough to accept them into the ontology, one last thing that can be done is to show *why* the 'bundle' alternative – x's identity is merely a conjunction of all of x's properties – is not satisfying. According to the 'bundle' view (a particular is *nothing but* a bundle of qualitative properties), individuation is ensured by the fact that no two particulars can be absolutely indistinguishable in the sense of possessing exactly the same set of qualitative properties. This claim has traditionally been expressed as the 'Principle of Identity of Indiscernibles' (PII): necessarily, any two particulars that have all the same qualitative properties are the same particular.[92] Of course, the plausibility of PII depends on what properties are to be included within its scope. That is why at least two forms of the Principle are commonly distinguished: PII(1) that excludes spatiotemporal properties from the domain of quantification (*strong version*), and (PII)(2) that quantifies over *all* properties without exception (*weak version*). Although many philosophers take PII(2) to be trivially true and, therefore, incapable of yielding interesting metaphysical theses (cf. Diekemper, 2009: 258, Russell & Whitehead, 1957: 57), it seems that counterexamples to both forms of PII can be provided. In the following lines, I will therefore introduce some classes of suitably arranged particulars for which either PII(1) or both PII(1) and PII(2) possibly fail to apply. This will serve to show that, since two particulars can share all their properties without being identical, there might be no better alternative than distinguishing them by accepting bare particulars in our ontology.

First, the most famous counterexample to PII(1) is arguably from spatial dispersion. Max Black's version of this counterexample, involving two iron globes, is the most commonly cited in the recent literature (cf. Black, 1952: 153–164).[93] We are to imagine a world consisting solely of two exactly resembling, large, solid globes of iron. These two globes always have been, are, and always will be exactly similar in shape (perfectly spherical), size, chemical composition, color, etc. And, importantly, it is not only their non-relational qualitative properties that these globes share, but also their relational properties. For instance, "[…] each of them has the property of being two diameters from another iron globe similar to itself" (Adams,

[92] PII should be contrasted with the uncontroversial Principle of the Indiscernibility of Identicals – necessarily, if two particulars are identical, then they have all the same qualitative properties.

[93] Other counterexamples to PII(1) from spatial dispersion are to be found in Kant ([1787] 1998: A263/B319) and Strawson ([1959] 2003: chapter 4 'Monads').

1979: 13).[94] In short, all the relations that one globe bears to the other are born by the latter to the former: it is a perfectly symmetrical world. In this example, the only reason why we know that these two globes are not identical is that they are spatially distant from one another, while the same particular cannot be in two places at once (i.e. a particular cannot be spatially distant from itself).[95] Since such a world seems to be logically possible, it must be concluded that there could be qualitatively indiscernible, yet numerically distinct, things, and therefore that PII(1) is false.

Second, taking the previous counterexample for granted, it is easy to set up a similar counterexample to PII(1) from temporal dispersion: we only have to substitute events for globes, and time for space. Assuming that events are non-repeatable, concrete particulars,[96] we can imagine a world consisting solely of temporally dispersed events, some of which being qualitatively indistinguishable. This world might consist of "[…] an infinite series of sounds … A B C D A B C D A …, succeeding one another at equal intervals, with no first or last term" (Hacking, 1975: 254). In this example, each term 'A', 'B', 'C', and 'D' refers to an event with a particular set of qualitative properties. For instance, perhaps every occurrence of A has the property 'having pitch A'.[97] It is worth elucidating this example with the type/token distinction: there are four types of events in this world, which are individuated by the set of qualitative properties (both relational and non-relational) that are instantiated by every token of that type. Thus, "[…] every occurrence of, for example, A, is a token of the event type 'A'; that is, an occurrence of an event with the properties 'having pitch A', 'being earlier than a token of event type 'B'', 'being later than a token of event type 'F'', etc." (Diekemper, 2009: 264). It therefore seems that each of the token events in this world shares all of its qualitative properties with all of the other tokens of its type. Yet, it clearly appears that tokens of event type 'A', for example, are all distinct events, given that they are temporally dispersed (a particular cannot be temporally distant from itself). Once again, since such a world is logically possible, it must be concluded that PII(1) is false.

Third, Steven French (2015) provides good reasons to believe that PII(1) fails in the quantum domain. As in the previous counterexamples, there indeed is a sense in which quantum objects of the same kind (such as electrons) – since they share all

[94] Some philosophers (e.g., Hacking (1975) argue that it should remain open to deny that such a world is correctly described as having two indiscernibles globes. After all, such a completely symmetrical situation of two globes could be re-interpreted as one globe in a non-Euclidean space. As Peter Forrest puts it, […] what might be described as a journey from one sphere to a qualitatively identical one 2 units apart could be redescribed as a journey around space back to the very same sphere" (2010: §3). So, to avoid question-begging one must have first to show that there are two globes on independent grounds. In reply, one may argue that, although it must be acknowledged that the very same world with indiscernible twins can be redescribed as a single-sphere world, this does make Black's description false. Both descriptions are true and this is enough to reject PII(1).

[95] This principle, though intuitive, might be doubted. For example, John O'Leary-Hawthorne (1995) argues that the bundle theory of substance can allow for identical, yet spatially dispersed, bundles of universals.

[96] Although this conception of events looks minimal, it is not free from controversy. It must be acknowledged that some philosophers, for example Roderick Chisholm, argue that events are repeatable and non-concrete entities.

[97] For the sake of simplicity, we will assume that the sounds and their properties are the sole constituents of these events.

their state-independent properties (charge, spin, rest, mass, etc.) – are indistinguishable. According to the 'Indistinguishability Postulate': "[i]f a particle permutation *P* is applied to any state function[98] for an assembly of particles, then there is no way of distinguishing the resulting permuted state function from the original unpermuted one by means of any observation at any time" (French, 2015: §2). So, if one wants to regard such particles as individuals (as Boltzmann (1887) urges to do), then we have to argue that their individuality resides in something over and above their state-independent properties (rejection of PII(1)).[99]

But it seems that quantum objects are indistinguishable in a much stronger sense: according to the most plausible understanding, "[…] no measurement whatsoever could in principle determine which one is which" (French, 2015: §4). Specifically, since some particles are taken to possess not only their state-independent properties in common, but also their state-dependent properties (i.e. "the properties expressed by expectation values of all quantum-mechanical physical magnitudes" (French & Redhead, 1988: 240)), then it can be shown that these particles violate PII(2) (which includes spatiotemporal properties). Indeed, whereas, in classical physics, the state-dependent properties of a particle are completely specified by the maximally specific state description, in quantum mechanics, there might be no pure states (i.e. no maximally specific assignment of expectation values) that can be ascribed to separate particles. Therefore, whereas, in classical mechanics, two particulars can always be distinguished via their spatiotemporal trajectories (since they cannot overlap – *impenetrability* assumption), the situation appears to be very different in quantum mechanics. Consider, for example, two fermions in a spherically-symmetric singlet state: they are not only indiscernible in the weak sense, but they also "[…] possess exactly the same set of spatiotemporal properties and relations" (French, 2015: §4, cf. also Ladyman & Ross, 2007: 135). It thus seems that if one wishes to maintain that quantum particles are individuals, then their individuality cannot be grounded without appealing to something like bare particulars, which implies that PII is false, even in its weakest form – PII(2).[100]

A last issue encountered by the 'bundle' view has to do with change. In brief, whereas the substratum theory makes it clear *how* a particular (e.g., Napoleon) undergoes a change (e.g., becoming the Emperor), the 'bundle' alternative has a hard time

[98] In quantum mechanics, the state function determines the probability of measurement results. Hence what the Indistinguishability Postulate expresses is that a particle permutation does not lead to any difference in the probabilities for measurement outcomes (cf. French, 2015: §2).

[99] Cf. also Huggett, 1999b.

[100] Of course, this conclusion has been challenged. First, some philosophers take quantum objects to be non-particulars in some sense, so that PII would not apply (cf. French, 2015: §4). Then, some philosophers have argued that each fermion in the above example enters into the symmetric but irreflexive relation of 'having opposite direction of each component of spin to …' on the basis of which they can be said to be 'weakly discernible' (cf. Saunders, 2003). This result was first established for fermions and then extended to bosons (and then generalized to infinite dimensional Hilbert spaces). However, as French and Krause (2006) have made clear, the appeal to irreflexive relations in order to ground the individuality of the objects which bear such relations involves *circularity*: two particles cannot be so related without having first been individuated. Finally, the non-orthodox Bohmian interpretation of quantum mechanics seems to offer room for a conception of quantum objects as particulars and for the preservation of PII (cf. French, 2015: §4, and Cushing et al., 1996).

to account for the same phenomenon. Roughly, the substratum theorist argues that when Napoleon becomes the Emperor, a thick particular (whose identity is independent of the properties it has) acquires the property of *being the Emperor*. Such a straightforward option is however not available to bundle theorists, since they take the particular 'Napoleon' to be nothing but the bundle (or the mereological fusion) of all his properties. As such, any property change (e.g., becoming the Emperor) seems to amount to a change in a bundle of properties and, therefore, to the creation of a new particular, rather than to a change in the same particular. In other words, the 'bundle' view seems to have the unwelcome consequence of construing change as "[…] a replacement of one individual by another, not change in the properties of one and the same individual" (Van Cleve, 1985: 98). Given the natural assumption that particulars remain identical (or survive) through changes, one should conclude that something is wrong with the 'bundle' view. For all these reasons, the substratum theory appears to be a much more plausible option than the 'bundle' view.

3.11 Conclusion

In the McTaggartian picture, GBT is introduced as an intermediary between two extreme answers – eternalism and presentism – that can be provided to the question 'Do the future and the past exist?'. However, this traditional way of introducing the A-theories of time is unsatisfying, since, from an ontological point of view, there is no pre-theoretic reason to adopt one theory rather than another. After all, the only events we experience are the ones occurring at the time of our experience, whether this be in eternalist or in presentist settings. Therefore, a proposal in this chapter was to make a fresh start in the debate by introducing the A-theories in terms of a new question: 'Is there a geometric asymmetry between the future and the past?'. The notion of 'geometry' refers here to basic, intrinsic properties of the spatiotemporal structures described by the A-theories. These structures are to be thought as ontologically prior to the individuals (e.g., spacetime points, events and objects) that participate within them (structuralism). In short, A-theories that possibly always answer 'no' to the above question are called '*symmetric*' (e.g., eternalism, presentism), whereas A-theories that necessarily sometimes answer 'yes' are called 'asymmetric' (e.g., GBT, SBT). One consequence of this proposal is that GBT is no longer to be seen as an ill-conceived hybrid between two polar opposites (eternalism and presentism), but as a real alternative to two *symmetric* theories.

However, the question 'Is there a geometric asymmetry between the future and the past?' was shown *insufficient* to distinguish between the various forms the *symmetric* and *asymmetric* theories may adopt. Therefore, in order to get a complete categorization of the various A-theories of time, a second question was introduced: 'Is temporal becoming (i.e. the creation of new things in the present) real?'. This second question allows one to distinguish between *pure becoming-views* (e.g., presentism, GBT) and *non-generative-views* (e.g., SBT, eternalism). In the McTaggartian picture revisited, GBT is thus singled out as the only *asymmetric* A-theory of time that accepts *Temporal Becoming*: $\mathbf{S}(\exists x\, \mathrm{H}\neg\exists y\; y = x)$. These new characterizations reveal GBT as being better positioned than its rivals to accommodate various past-future time asymmetries, including the asymmetry between the 'open future' and the 'fixed past'. First, since GBT is an *asymmetric* A-theory, it can avail itself of the 'no fact of the matter' account of the openness of the future, while keeping the past fixed: unlike presentism, GBT allows one to say that facts about what happened *supervene* on the spatiotemporal structure, while facts about what will happen do not. Second, assuming that *physical determinism* is false, GBT implies, through *Temporal Becoming*, that new things are created in the present, while (at least) some of them are not made inevitable by how things located in the present or the past of now are or were. GBT thus satisfies the two necessary conditions (introduced in the second chapter) to capture a great variety of senses in which the future can be called 'open', including the most radical ones.

Despite some attractive features, GBT is often criticized for not being a *viable* alternative to presentism because of the 'epistemic objection'. According to this objection, GBT would provide no basis for saying that our time is the objective present. Worse, this theory would commit to conclude that we are, almost certainly, located in the objective past. However, as has been argued, this objection relies on a mistaken assumption, namely that the reality of the past entails that events are occurring in the past. Indeed, it is one thing to say that the past is real, it another to say that past events are still occurring. In particular, assuming that bare particulars of past events will always exist, GBT can be reconciled with the intuitive idea that events can only occur at the present time. In this perspective, becoming past involves alterations in a thing's intrinsic properties to such an extent that it ceases to belong to its *natural kind*, while the bare particular of *that* thing will continue to exist. The intrinsic properties that are lost in the process are (at least) those that define the natural kind to which the thing in question belonged when present. As a result, since (i) there are no events – such as 'Napoleon's thinking about crowning the Emperor in the present' – occurring in the past, and (ii) if one's belief is occurring, then one introspectively knows it, we (as constituents of conscious events) can be confident we are located in the present, i.e. at the leading edge of the growing block. The above story about *how* things located in the past exist, involving bare particulars, is not *ad hoc*, since (i) bare particulars are useful in various independent contexts (e.g., temporal ontology, metaphysics of spacetime and metaphysics of mathematical entities), and (ii) the 'bundle' alternative is not satisfying (e.g., there are classes of individuals for which both weak and strong versions of the Principle of the Identity of Indiscernibles fails to apply.

Chapter 4
Reconciling the Asymmetry with Contemporary Physics

Abstract As has previously been argued, the growing block theory of time (GBT), since it is essentially *asymmetrical* (necessarily sometimes the structure it describes is not reflection invariant), while it accepts *Temporal Becoming* (new things are created in the present), is better positioned than the traditional models of the temporal structure of the world (eternalism and presentism) to accommodate our intuition that the future is open and the past fixed. However, GBT (like any other A-theory of time) is often criticized for conflicting with some important results of contemporary physics (e.g., by requiring an absolute notion of objective simultaneity). In this chapter, I argue that GBT, far from being disqualified by contemporary physics, might be underpinned by some recent approaches to quantum gravity, especially CST.

4.1 Introduction

At first sight, GBT appears at odds with some important results of contemporary physics. In particular, since GBT is usually based on the idea that the layers of existence successively coming into being are slices of Newtonian absolute time, this theory could hardly be reconciled with the developments of the Special and General theory of relativity. For example, Hilary Putnam (1967) and Wim Rietdijk (1966) have both argued that, since the view that the future is unreal requires an objective notion of absolute simultaneity, it is incompatible with the Special theory of relativity (SR), according to which the simultaneity of space-like separated events is relative. However, not only the assumptions of these arguments can be disputed (cf. Bourne, 2006; Correia & Rosenkranz, 2018; Miller, 2013; Sklar, 1974; Stein, 1968, 1991; Tooley, 1997; Zimmerman, 2008), but recent approaches to quantum gravity have been formulated in terms that echo C. D. Broad's theory. For example, Raphael Sorkin put forward a model of causal set dynamics according to which "[…] reality is more naturally seen as a 'growing being' than as a 'static thing'" (2007: 153). In other words, even if GBT turns out to be inexpressible in a relativistic framework (which is doubtful), one can question the credentials of the theories of relativity

V. Grandjean, *The Asymmetric Nature of Time*, Synthese Library 468,
https://doi.org/10.1007/978-3-031-09763-8_4

(especially given their incompatibility with quantum mechanics), and argue that our most fundamental physics is rather to be found in the nascent theories of quantum gravity, while some of them seem compatible with GBT.

Specifically, the causal set approach to quantum gravity (CST), which dismisses the spacetime continuum as mere approximation[1] in favor of locally finite causal sets, seems consistent with C. D. Broad's notion of temporal becoming introduced in the previous chapter. Temporal becoming could indeed be restored by a dynamics, called the 'classical sequential growth dynamics' (CSG), by which the growth in the causal sets takes place. Thus, although the search for a quantum theory of gravity is currently dominated by String Theory and Loop Quantum Gravity, CST might help solve some of the key problems encountered in trying to make GBT work in a relativistic setting. Of course, CST – or any other scientific research program – cannot settle the debate about the nature of time, but it can at least *inform* and *frame* the debate, so that metaphysicians will be offered new ideas and be prevented from saying too much nonsense.

This chapter is structured as follows. In the second section, I briefly introduce some basic elements of (neo-)Newtonian physics and show *why* it is a friendly environment for GBT. In the third section, I present the 'relativity revolution' that emerged out of Albert Einstein's observations, by introducing the main postulates and consequences of the Special theory of relativity (SR); I lay particular emphasis on Minkowski's attempt to make sense of SR as a *spacetime* theory. In the fourth section, I expose the threat that SR may represent for GBT, especially through its rejection of an objective notion of absolute simultaneity, and different strategies (both incompatibilist and compatibilist) that growing blockers may adopt to escape this threat. In the fifth section, I go beyond SR's limited aspirations by introducing the General theory of relativity (GR) and consider some of the main difficulties in squaring it with quantum mechanics (especially with regard to *causal relativity*). In the sixth section, I expose one approach to quantum gravity which, since it encompasses both relativistic and quantum phenomena, aims to address these difficulties. Moreover, I show that this approach, the causal set approach, promises to provide a notion of temporal becoming, and could thereby make GBT an attractive model in the contemporary scientific context. I conclude against a widespread opinion that fundamental physics does not undermine any attempt to defend an open-future view. Finally, in the seventh section, I move from science to science fiction and show that GBT is, in principle, compatible with some scenarios such as time-travel.

[1] The term 'approximation', although it is common in physics, is misleading, since the continuum is more fine-grained than a discrete structure. The idea is rather that the continuum view models spacetime in a way that is generally regarded to be more convenient (for both mathematical and physical reasons) than the discrete view.

4.2 (Neo-)Newtonian Basics

For more than two centuries, physicists have assumed, in accordance with Isaac Newton's *Principia* (1687), that space was an infinite and immutable three-dimensional Euclidean continuum of points, all of which persist through time, which is also infinite.[2] The Newtonian view can be described, somewhat anachronistically, in terms of *spacetime*, provided that persisting spatial points are replaced with a succession of numerically distinct spacetime points. Newtonian spacetime is then to be conceived as a four-dimensional continuum consisting of a stack of three-dimensional volumes of space (or *hyperplanes*), each instantaneous and spread out continuously in a fourth, temporal, dimension. Intuitively, if one thinks of Newtonian space as a cube, then one should think of Newtonian spacetime as a succession of cubes. It is however often useful to pretend that space has just two dimensions, so that we can visualize Newtonian spacetime as a succession of flat-surfaces, each of them representing a three-dimensional hyperplane of *absolutely* simultaneous spacetime points (cf. Fig. 4.1). In such a diagram, there are spatial and temporal distance relations between every point (including spatial distance relations between every point in different hyperplanes). Furthermore, each persisting object is represented by a *worldline* (or a *worldtube*, when the object is not a point), which depicts its complete trajectory through a succession of different spacetime points. Specifically, Newton's laws of motion[3] allow for the following description, which refers to Fig. 4.1: objects that are at absolute rest (*a*) have vertical straight worldlines; objects in uniform motion (*b*) have non-vertical straight worldlines that cut through the flat-surfaces (the greater the degree of deviation from the vertical, the faster the velocity); and finally objects that undergo acceleration (*c*) have curved worldlines (the steeper the curve, the greater the acceleration – this is a little loose, since a motion could actually be such that acceleration could be greatest at point where tangent to curve is vertical).

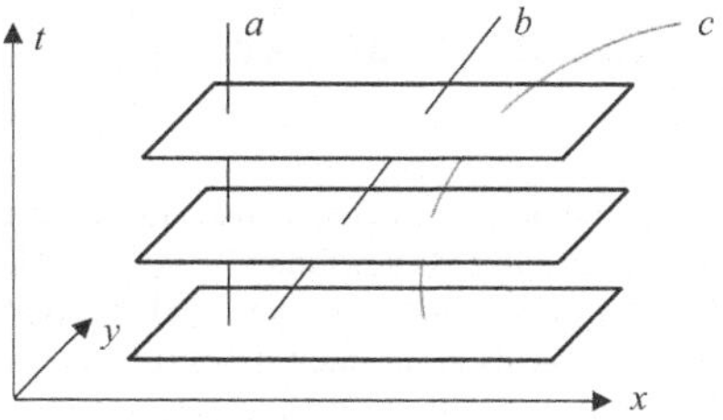

Fig. 4.1 Newtonian spacetime

[2] This conception was, for instance, made explicit in Newton's essay *De Gravitatione* (cf. Huggett, 1999a: 113).

[3] Newton's first law of motion simply states that: "[…] any body not subject to a force travels along a straight path through space-time" (Maudlin, 2011: 35). By contrast, a body subject to a force will occupy a curved trajectory, and Newton's second law states exactly how the path will curve (cf. Maudlin, 2012: 59–60).

It is worth mentioning that if one abolishes the spatial distance relations between points in different hyperplanes, while leaving everything else unchanged, one turns a Newtonian spacetime into a neo-Newtonian (or Galilean) spacetime.[4] This seems important, since the distinction between absolute rest and uniform motion, which is obtain through *these* cross-temporal relations, appears to be superfluous. After all, Newtonian physics itself predicts that no experiment whatsoever could determine the absolute rest of an object: "[l]aboratory experiments done in a room at absolute rest would have identical outcomes to those done in a room moving with constant velocity" (Maudlin, 2011: 36). Thus, whereas a neo-Newtonian spacetime preserves the distinction between straight and curved trajectories, it abandons cross-temporal spatial distances: only points that are simultaneous are located at a spatial distance from one another. It is therefore wrong to think of classical physics as entirely free from relativity. In a neo-Newtonian spacetime, for instance, statements such as ⌜This object is moving has a speed of 10 km h^{-1}⌝ or ⌜This object is at rest⌝ are meaningless; people in different frames of reference[5] will disagree about what speeds objects have, and there is no objective fact of the matter as to which groups are right. Most interesting differences and similarities of Newtonian and neo-Newtonian spacetimes are summarized in the Fig. 4.2.

Newtonian and neo-Newtonian spacetimes show that, although the word 'spacetime' is spontaneously associated with the 'block universe' view of time, it is perfectly compatible with *dynamic* conceptions of time.[6] In particular, since both Newtonian and neo-Newtonian spacetimes allow for an *objective* partition (or *foliation*) into slices of events that all happen *at the same time*, it seems that a dynamic theory such as GBT can fairly be expressed within classical physics. To put it another way, although classical physics does not postulate a *now*, it allows for the possibility of an objective spatially extended present, and even for temporal becoming. For example, John Earman (2008) argues that, since temporal becoming consists of the infinite accretion of layers of existence, which are naturally seen as slices of Newtonian absolute time, it can be implemented in the Newtonian setting, which thereby offers a growing model of the universe. Roughly, Earman's idea is to define a model of GBT as a pair $\{\mathfrak{M}, \precsim\}$, where $\mathfrak{M}$ is a set of spacetime models, each of

---

[4] I am simplifying matters considerably. Actually, neo-Newtonian spacetime has 4-D affine structure. In context of Newtonian spacetime (despite standard presentations, such as Earman 1989), this is not an autonomous element, but is provided by the facts of the matter about spatial distances between points at different times (together with the temporal metric). Once one gets rid of those one needs to 'add' the affine structure as a new primitive element.

[5] As used herein, the term 'reference frame' is an observational perspective (that is established by centering a coordinate system on a particular body that one assumes to be at rest) from which the motions of all other bodies (relative to the reference body) can be specified.

[6] Those, such as Maudlin (2002), who defend the 'block universe' view as the basis for an explanation of some of our intuitions about time might feel unfairly stigmatized by the 'static/dynamic' labels. But these labels have a specific meaning in the present context: a theory is called 'dynamic' iff it accepts either *Temporal Becoming* or *Annihilation* (or both) (e.g., presentism, MSL, GBT, SBT); conversely, a theory is called 'static' iff it accepts neither *Temporal Becoming* nor *Annihilation* (e.g., B-eternalism) (cf. Sect. 3.3).

	Newtonian spacetimes	Neo-Newtonian spacetimes
Spatial and temporal invariant distances between any two points	Yes	No (there is no spatial distance between points at different times)
Distinction between straight and curved paths through spacetime	Yes	Yes
Distinction between absolute rest and absolute motion	Yes	No
Absolute velocities	Yes	No

Fig. 4.2 Comparison between Newtonian and Neo-Newtonian spacetime

which is isomorphic to a 'future truncated' version of a Newtonian spacetime, i.e. to a model that results from deleting from a Newtonian spacetime all points later than some particular time, and restricting the geometric and physical fields to that truncated spacetime. The relation $\lesssim$, which expresses temporal becoming, holds between two of the spacetime models in $\mathfrak{N}$ iff one can be isomorphically embedded as a submodel of the other.[7] Thus, although expressing GBT in Newtonian setting requires some efforts, classical physics (with absolute simultaneity) appears to be a friendly environment: (at least) some versions of GBT seem respectful of its main imperatives.

However, at the beginning of the twentieth century, physics underwent a deep revolution, known as the 'relativity revolution', which entailed the rejection of the classical theory – though it is still considered as an accurate approximation at low velocities (relative to the speed of light). This revolution, which emerged out of Albert Einstein's observation that the classical theory conflicts with Maxwell's equations of electromagnetism and, experimentally, with the Michelson-Morley outcome,[8] seems to exert significant pressure on some philosophical theories of time, and especially on GBT. In particular, both the Special and General theory of relativity rule out the Newtonian objective notion of absolute simultaneity, which seems required in the definition of GBT. According to both SR and GR, there is simply no objective fact of the matter as to whether two space-like separated events are simultaneous or not. The Putnam-Rietdijk argument is probably the most famous argument pointing in this direction. However, before detailing this argument and

[7]As Natalja Deng (2017) points out, although Earman's version of GBT is formally elegant, it meets difficulties when it comes to interpretation. For example, it seems that, in such a model, the statement 'The future does not exist' is merely *perspectivally* true. That is, it is only from the perspective of each of the spacetimes models of $\mathfrak{N}$ that the future is not real; otherwise there would be no growth. This clearly seems to betray C. D. Broad's intention. However, the point here is not to show that formulating GBT within Newtonian setting is straightforward, but merely that problems will get worse in Special and General relativistic spacetimes.

[8]To be historically accurate: the Michelson-Morley outcome is not mentioned in Einstein's, 1905 paper.

possible options to answer it, it is worth recalling what the relativity revolution is about. In the following pages, I will therefore introduce the main postulates and consequences of Einstein's early work.

4.3 The Relativity Revolution

In 1873, James Clerk Maxwell published a work, entitled *A Treatise on Electricity and Magnetism*, in which he formulated *the classical theory of electromagnetic radiation*, bringing together electricity, magnetism, and light as different manifestations of the same phenomenon. Specifically, Maxwell's equations of electromagnetism, which have been called the 'second great unification in physics' (after the first one realized by Newton), show *how* electrical and magnetic forces are intimately related, and especially *how* electrical and magnetic fields could combine to form self-propagating *electromagnetic waves*. The velocity of such waves, which can be predicted from the equations, turned out to be the speed of light: $c = 300{,}000$ km s^{-1}. This result led Maxwell to suppose that light itself was an electromagnetic wave (which turned out to be the case). Since waves had always been studied as perturbations in something (e.g., water, air, the surface of drums, etc.), physicists immediately assumed that light waves also consisted of perturbations in a medium: *luminiferous aether*. This was the advent of 'aether-physics', which was in charge of capturing the properties of this invisible, but all-pervasive medium. It was, for instance, commonly agreed that aether must be in a state of absolute rest. This consensus was to prove fruitful, since it opened the possibility of experimental tests for distinguishing absolute motion from absolute rest (cf. Dainton, 2010: 315-317).

Such an experimental test was first carried out in 1881 by Albert Michelson alone (and then in collaboration with Edward Morley in 1887), but its outcome defied the odds. Basically, the idea was to compare the speed of light in perpendicular directions, in attempt to detect variations through the absolute stationary aether. After all, given that waves are not affected by how fast their sources are moving (contrary to projectiles),[9] if light rays travel in all directions at $c = 300{,}000$ kms^{-1} (as Maxwell's equations predict), this speed should only be obtained when the test is conducted at rest relative to the aether. If the speed of light is measured by someone who is moving through the aether, he should find that the light rays travelling in the same direction as him are moving *slower* than those travelling in the opposite direction, or so theory predicted. Thus, assuming that the Earth moves through the aether as it orbits the sun, Michelson measured the time it took for light to travel along two paths of equal distance; one of these paths was aligned in the direction of the Earth's motion around the sun, the other at right angles to this direction. But, whereas Michelson expected the light ray travelling parallel to the flow of the aether to be

[9] For example, whereas the velocity of a cannonball depends on the velocity of the cannon that expels it, the sound waves propagated by a plane do not move faster in the direction the plane is flying in.

slower than the one travelling perpendicular to it, no variations in the speed of light (through the presumed aether) were detected. Light turned out to *always* travel at the same speed, irrespective of the Earth's motion.[10] This could not be explained by classical physics (especially because of the Galilean transformations),[11] which therefore had to be (substantially) revised.

In that respect, Einstein's radical proposal,[12] called 'Special Relativity' (SR), results from the willingness to follow through to the end the consequences of two postulates: (i) *the Relativity Postulate*, according to which the laws of nature do not distinguish between different observers undergoing inertial (i.e. force-free) motion, and (ii) *the Light Postulate*, according to which the numerical value of the speed of light is the same when measured in any direction, by any inertial observer.[13][14] This proposition involves a reformulation of mechanics in terms of the *Lorentz transformations* (instead of the *Galilean transformations*) that reflect the fact that observers moving at different velocities may measure different distances, elapsed times, and even different ordering of events, but always such that the speed of light is the same in all inertial frames of reference. In short, one could say that the relativity revolution consists in replacing one absolute, Newton's absolute space and time, with another, the velocity of light. This famously requires the notions of distance and time (both involved in the definition of velocity) to vary in systematic ways in the relevant frames. It is precisely these variations, sometimes misleadingly called 'length contractions' and 'time dilations', that the Lorentz transformations allow one to calculate.[15] Surely the most natural comparison here is the absolute structure of Minkowski spacetime, which postulates a different pattern of spatial and temporal distances to (neo-)Newtonian spacetime. If one takes the light speed as the absolute, one could consider Harvey Brown's *Fable of Keinstein* (2005: chap. 3), and see forces and masses as the absolutes that allow one to derive the Galilean transformations in a way analogous to Einstein's derivation of the Lorentz transformations.

---

[10] This affirmation is true, provided that light travels in vacuum. If light travels through a medium such as water or air, it will be slowed down.

[11] Galilean transformations, which are a set of equations relating one set of coordinates to another, suppose that the distances and times between objects are absolute.

[12] Of course, Einstein's proposition was not the only one on the market. See, for instance, Fitzgerald's *compensatory theory* (cf. Dainton, 2010: 317).

[13] Strictly, this is not Einstein's *Light Postulate* (which is a claim about light in the "stationary" frame) but an elementary consequence (the first Einstein drew) of that postulate when combined with the relativity principle.

[14] The acceptance of these two postulates *presupposes* the rejection of the notion of *luminiferous aether* and with it the notions of absolute motion and rest. After all, if the velocity of light is unaffected by the Earth's motion through the aether (as the Michelson-Morley outcome stated), physicists are probably wrong in supposing that the aether exists.

[15] The notions of 'length contractions' and 'time dilations' are misleading, because they suggest that a body, at any given moment, has a speed that can be closer or further from the speed of light. But, just as in neo-Newtonian spacetime, there simply are no such speeds: there is, for instance, no fact about how fast the Earth is moving right now (cf. Maudlin, 2012: 68).

To get a feel for what this means in practice, imagine two people, Max and Mary, travelling outer space. Suppose that Max switches on a torch and sends a light signal in Mary's direction; he therefore sees a light ray moving at $c = 300{,}000 \text{ km s}^{-1}$ towards her. Furthermore, suppose that Max sees Mary moving away from him at the very high speed of $0.5c$; if she measures the speed at which Max's light signal passes her by using a clock, she will also find that it is moving at c. Indeed, given *the Light Postulate*, the speed of light is constant in all frames of reference, and independently on the motion of the source, which implies that moving away from its source does not make it appear greater. How can this story be possible? Roughly, Einstein's answer is this: if Max could look at Mary's clock, he would find that it is running *slower* than his. In other words, keeping the speed of light constant for Max and Mary requires their clocks to be *desynchronized*: a clock that is moving relative to an inertial frame of reference will be measured to tick slower than a clock that is at rest in *this* frame of reference. But, given *the Relativity Postulate* and the abandonment of absolute motion, Einstein goes a step further by arguing that each can regard themselves as at rest with as much right as the other. Counterintuitively, that means that if Mary could look at Max's clock, she would *also* find that it is running slower than hers.[16] Thus, although Max and Mary's readings are different, Einstein argues that they are *equally valid*; there is no objective fact of the matter as to which of their clocks is accurate.[17] This example is inspired by Dainton (2010: 317).

According to its standard interpretation, SR has many weird consequences; some of them directly impact the classical A- and B-theories of time, especially GBT. I will mention 3 of the most striking consequences. First, there is no objective fact of the matter about which events happen at the same time. Whether or not two events are simultaneous depends on the frame of reference from which they are considered; observers in different frames of reference will find different events simultaneous, and there is no sense in saying that one observer is right and another wrong. Second, space and time are not to be conceived as two separable and quite distinct entities, but much rather as entangled aspects of the same underlying four-dimensional continuum that fuses the two into a spacetime, the so-called 'Minkowski spacetime' (cf. 'unitism' contrasted with 'separatism' in Gilmore et al. (2016)). Although we are already familiar with the general idea of spacetime (cf. Sect. 4.2), Minkowski spacetime interestingly differs from the two types of structure we have encountered so far, Newtonian and neo-Newtonian, especially with respect to the quantities it takes to be invariant: neither spatial nor temporal distances are

[16] More generally, SR predicts that when two observers are in motion relative to each other, each will measure the other's clock slowing down, in concordance with them moving relative to the observer's frame of reference. This prediction goes against common sense, according to which if the passage of time has slowed for a moving observer, this observer would observe the external world's time to be correspondingly sped up. However, no contradiction is to be found here. Similar effects of perspective can be observed in everyday life: if two persons, A and B, observe each other from a distance, B will appear smaller to A, but at the same time, A will appear smaller to B.

[17] Actually, from the Minkowskian point of view, both clocks can be accurate – they both reveal the objective temporal distances along their own trajectories.

invariant, the speed of light and the spacetime interval are.[18] Third, considering two twins, Max and Mary, if Mary departs on a space journey and subsequently returns, she will be younger than stay-at-home Max. Time has passed at a slower rate for her than for Max (which appears at odds with *the Relativity Postulate*). This is due to the geometry of Minkowski spacetime (which will be introduced below): if one calculates the length of Mary's trajectory in a Minkowski diagram, one will find out that it is smaller than the length of Max's trajectory, even though Euclidean representations of the case present Mary's path as being *longer* than Max's path (which explains why the case may seem paradoxical).[19]

At first sight, one might be surprised to find out that two seemingly innocuous postulates – *the Relativity Postulate* and *the Light Postulate* – have such enormous consequences. But, one has to remember that the difficult step, which Einstein first took, was to consider seriously the possibility that these postulates *might* be true at a time when this was far from obvious. Among these consequences, the relativity of simultaneity, which undermines the Newtonian assumption that there is an objective fact of the matter as to which events occur at the same time, is presumably the most problematic for GBT.[20] It is therefore worth considering it in greater detail. Basically, the idea behind the relativity of simultaneity is that whereas there is no problem in establishing a common time system for a group of people who are stationary with respect to one another, so that everyone in this group will agree on which events are simultaneous, things get harder for people who are *moving* with respect to one another: it is *impossible* for them to synchronize their clocks.[21] Examples involving high-speed trains and lightnings are commonly used to illustrate the situation.

For instance, Einstein (1920: 25–27) invites us to consider a train, both ends of which are struck by bolts of lightning, producing two flashes. Whether or not these two flashes occur *simultaneously* depends on the motion of the observers relative to the location of the events. For example, if Max stands by the side of the track and, when the lightning strikes, is at the mid-point of the high-speed train, he will see the two flashes as occurring *simultaneously*. However, if Mary is sitting midway inside the train when the lighting strikes, she will see the two flashes as occurring *successively*. In other words, the two flashes have different time coordinates in frames of

[18]Actually, this is sub-optimal, since the very notion of a speed is a frame relative one. Speeds do not exist in Minkowski spacetime unless and until one picks out a frame of reference (i.e. one adopts some standard, an arbitrary choice between sets of coming inertial trajectories) and defines in terms of it a coordinate time, effectively persisting space, etc. Minkowski spacetime does have frame-independent (which is not the same as frame-invariant) spatial and temporal distances: for any everywhere spacelike or everywhere timeline curve in spacetime, there is an absolute fact of the matter about its spatial or temporal length.

[19]For a detailed discussion on this consequence, called the 'Twins Paradox', see Maudlin, 2012: 77–83.

[20]In comparison, the second consequence is no big deal for GBT, since this theory describes time as much an objective dimension of reality as any of the three spatial ones (cf. Grandjean, 2022).

[21]Things will turn out to be more complex in GR since, loosely speaking, not only velocities but also gravitational potential differences between locations may affect which events are found to occur simultaneously.

reference that are in motion relative to each other.[22] From Max's perspective, Mary's deviant observation can easily be explained by the fact that she is moving towards the light travelling from the front of the train, and she thus *decreases* the distance the light has to travel to reach her. But, given *the Relativity Postulate*, Mary's perspective, which takes the ground to be moving beneath the stationary train, is not less valid than Max's one. She is therefore perfectly legitimate in saying that it is Max's observation that is deviant, because he *increases* the distance between him and the light coming from the front of the train. We are therefore left with two flashes, the time ordering of which differs depending on who observes them, and there is no objective fact of the matter as to which observer is right.

It may be helpful to visualize the situation using Minkowski diagrams, which aim to graphically depict a portion of the new conception of spacetime that seems implied by SR. In such two-dimensional diagrams, where space has been curtailed to a single dimension, the vertical axis t (or t') refers to temporal and the horizontal axis x (or x') to spatial coordinate values. The introduction of a separate x' axis is required by the Lorentz transformations, according to which observers moving at different velocities may measure different ordering of events. Intuitively, the lines parallel to x (or x') correspond to the usual notion of *simultaneous events* for a stationary observer. Minkowski diagrams thus allow a qualitative understanding of the relativity of simultaneity (and other relativistic phenomena) without mathematical equations: each observer interprets all events on a line parallel to his x (or x') axis as simultaneous. Specifically, considering the previous case, a Minkowski diagram (cf. Fig. 4.3) allows one to show that whereas the two flashes occur simultaneously in Max's perspective (since they are both situated on x), they occur *successively* in Mary's perspective (since they are not situated on a line parallel to x'). But, in both perspectives, the two flashes are spatially separated (even though their spatial distance is also frame-relative), which corresponds to the front and the back of the train.[23]

It is worth saying a bit more about the standard way of construing Minkowski's attempt to make sense of SR as a theory of *spacetime*. Unlike Newtonian and neo-Newtonian structures, which both take some spatial and temporal distances as invariant (e.g., the spatial distance between points that occur at the same time), Minkowski spacetime is built around another invariant quantity: the speed of light. Indeed, since neither spatial nor temporal distances are invariant in SR, but the speed of light is, Minkowski's proposal is to integrate it into the very structure of spacetime. This seems to be a natural response to the abolition of *luminiferous*

[22] It is worth noticing that, given their position and the absoluteness of the speed of light, Max and Mary are both legitimate in their observations.

[23] Although two observers in different frames of reference may disagree about the spatial and temporal distance between events, they will agree on the spacetime interval between them. The relevant formula to calculate a spacetime interval I is: $I^2 = d^2 - c^2t^2$ (where d is the spatial distance between events, t is the time separation, and c stands for the speed of light). Thus, the speed of light is not the only invariant in Minkowski spacetime; the interval between points and events is the same in all inertial frames of reference (cf. Dainton, 2010: 323).

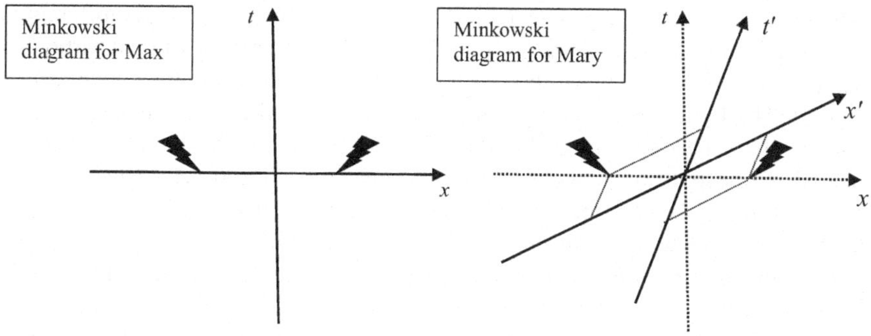

Fig. 4.3 Two Minkowski diagrams

Fig. 4.4 The light-cone structure

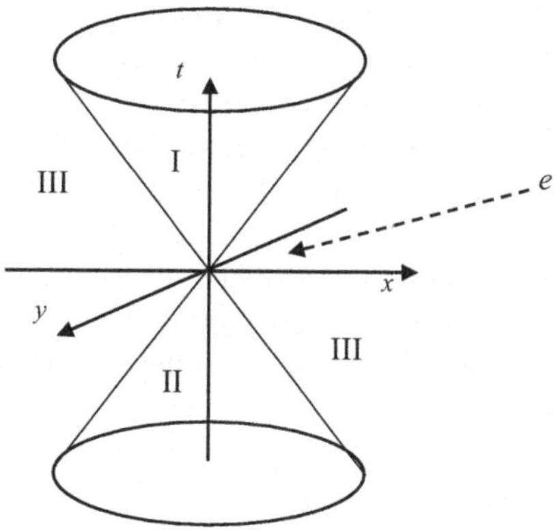

aether: given that light is no perturbation within any medium whatsoever, it seems that the only candidate available for determining the paths light rays can take is spacetime itself. The notion of a *light-cone* is crucial to understand the luminal structure of Minkowski spacetime. A light-cone is all the possible light rays that can be sent from (and received by) a single event *e* (occupying a certain spacetime point), which together trace out a double cone. Specifically, if one imagines light confined to a two-dimensional plane, the light rays emanating from *e* spread out in a spherical surface; and if one depicts this expanding spherical surface with the vertical 'time' axis, the result is a cone, known as the future light-cone. The past light-cone behaves like the future light-cone in reverse: a spherical surface contracts in radius at the speed of light until it converges to the spacetime point occupied by *e*. When the future light-cone (representing light emitted by *e*) is combined with the past light-cone (representing light received by *e*), they together yield the characteristic hourglass shape shown in the Fig. 4.4.

Given that light has the maximum possible speed (this can be derived from *the Light Postulate* and the Lorentz formula for velocity addition), the light-cone of *e* partitions the remainder of the universe into 3 separate (but topologically connected) regions (as indicated in Fig. 4.4): region I is known as the 'absolute future', region II as the 'absolute past', and region III as the 'absolute elsewhere' of *e*. In particular, the spacetime points lying *on* the surface of regions I and II are those that can be connected by light rays travelling in a vacuum, and are called 'light-like separated' from *e* – their spacetime interval is zero. The spacetime points lying *inside* regions I and II are those that can be connected by signals travelling *slower* than light, and are called 'time-like separated' from *e* – their spacetime interval is positive. Finally, the spacetime points lying *in* region III (i.e. outside the light-cone) are such that only a signal travelling *faster* than light could connect them; they are called 'space-like separated' from *e* – their spacetime interval is imaginary. Events that are very close together in time but spatially at a significant distance typically fall into this third region. It is worth mentioning that, for any space-like separated events, there is an inertial frame in which they are simultaneous, whereas time-like separated events are *non*-simultaneous in all inertial frames (though there are frames in which they occur at the same *spatial* coordinates). Thus, just as has been shown that classical physics is not entirely free from relativity, one finds that the relativity of simultaneity has limits within SR: relativization applies only to space-like separated events.

To illustrate, consider event *e*, which lies where the tips of the two cones meet (as indicated in Fig. 4.4). In its frame of reference, *e* is simultaneous with all the space-like separated events that compose a horizontal hyperplane (parallel to the *x*), slicing through the absolute elsewhere (region III). Yet one can imagine alternative hyperplanes centered on *e*, slicing through the absolute elsewhere *at different angles*. What I am assuming is a timeline direction at *e* that can be thought of as encoding the instantaneous state at *e* of an observer who has *e* on their worldline (e.g., the *t* axis in the diagram). Specifically, each of the alternative hyperplanes contains events that, from *e*'s frame of reference, lie either in the past or in the future, but *not* in the *absolute* past or future. Furthermore, all the events that are space-like separated from *e*, i.e. all the events that are in the absolute elsewhere of *e*, are simultaneous with *e* from some inertial frame. But there is no frame of reference in which events that are time-like separated from *e*, i.e. events that are in *e*'s *absolute* past or future, are simultaneous with *e*. Does that mean that, from the perspective of each spacetime point, the temporal ordering of *half* the entire universe is frame-relative? Not necessarily; "[…] the bulk of spacetime may well lie in our absolute past and future" (Dainton, 2010: 326). Indeed, one has to remember that indicating light-like connections with forty-five-degree lines results from a useful, but potentially misleading conventional choice.

Finally, it is worth mentioning that, since no physical influence can travel faster than light, light-cones allow for a *causal* interpretation: events that are time-like separated can be causally related, but no causal relation can hold between space-like events. To put it another way, whereas no event in region III can be the cause or the effect of *e* (since any influence would have had to travel *faster* than light), any event in region II can, in principle, influence (or be the cause of) *e*, and any event in region I can, in principle, be influenced (or be the effect of) *e*. Accordingly, regions I and II are sometimes renamed the 'causal future' and the 'causal past' of *e*. For example, two planets can causally interact only if they are linked by paths through spacetime that are always time-like (i.e. paths that always stay *inside* the relevant light-cones). If the spacetime paths between the two planets are space-like (even only partially), they cannot causally interact. Therefore, although the temporal ordering of events is affected by relativity, this cannot be extended to causal ordering: the relativity of simultaneity only applies to events that nothing travelling slower than light, such as causal influence, can connect. That is *why* Minkowski spacetime is often said to embody the causal structure of the universe.[24]

In a nutshell, unlike Newtonian and neo-Newtonian spacetimes, Minkowski spacetime, which seems implied by SR, admits no *uniquely correct* foliation into three-dimensional simultaneity slices (or *hyperplanes*); the objective notion of absolutely simultaneous events is therefore meaningless in the relativistic context. In other words, whereas the (neo-)Newtonian approach assumes that absolute simultaneity of two events could be established by suitable measurements, the Einsteinian approach assumes that, since measurements of the time elapsed between two events rely on light, which has a finite speed, no measurement whatsoever could establish an objective notion of absolute simultaneity between them. In relativistic physics, the light-cone structure has thus replaced the objective foliation. It is important to keep in mind that the light-cone structure allows for a causal interpretation (events that can causally interact[25] with e are only located in regions I and II); this interpretation will remain true in GR (Sect. 4.5) and will play a crucial role in the causal set approach to quantum gravity (Sect. 4.6).

4.4 Relativity as a Threat to GBT

Given what has been said so far, it could be argued that SR exerts some pressure on GBT (and on other classical A- and B-theories of time), at least as defined in the previous chapter. Such an argument has originally been developed by Hilary Putnam (1967) and Wim Rietdijk (1966) who have famously concluded that the view that the future is unreal is *incompatible* with SR. Contrary to what one might think, this conclusion is not derived from the mere fact that SR is a spacetime theory, although past and future light-cones may give the impression that Minkowski spacetime is an eternal block of events. As has been shown, (neo-)Newtonian physics can also be formulated as a spacetime theory, while nothing *a priori* prevents from implementing distinctions between the future and the past in the (neo-)Newtonian setting (cf. Earman, 2008). More threatening for GBT is what SR has to say about the present

[24] However, as will be shown in Sect. 4.5, the requirement that causal processes or signals can propagate only within the light-cone, is sometimes violated in the quantum context (cf. *causal relativity*).

[25] The term "interact" is however a little misleading, since it suggests influence of A on B and of B on A. This means that, strictly speaking, things that can be said to interact cannot be (point-like) events like e; they must persist/extend through time so that a stage of the first can affect a stage of the second and vice versa.

itself. According to SR, the distinction between what is present and non-present has no ontological significance, but depends on an arbitrary choice of frame of reference, and the same seems to apply to the distinction between what is real and unreal.[26]

Specifically, the Putnam-Rietdijk argument establishes that an observer *here-now* in Minkowski spacetime must, on pain of inconsistency with his base conventions, assign the label 'real' to events in the future light-cone of *here-now* (cf. also Savitt, 2000, Petkov, 2006, Dorato, 2008, Earman, 2008, Dainton, 2010, Wüthrich, 2010, Miller, 2013).[27] Roughly, the reason is that events in an observer's future light-cone can be in another observer's *relative* past or present, at a stage when the first observer judges the second observation to be real. This is an immediate consequence of the relativity of simultaneity: two observers in relative inertial motion (e.g., Max and Mary in the train example) will disagree about whether some set of events occur at the same time or not, and there is no sense in saying that one observer is right and another wrong. To put it another way: relative to inertially moving observers, spacetime will foliate differently into three-dimensional space-like hyperplanes, so that different sets of events will be simultaneous. Whereas this is no big deal for the B-theorist of time, since she does not ontologically discriminate between regions of spacetime,[28] things are more complicated for the growing block theorist (and other A-theorists). This can be illustrated by the following example.

Suppose that Max and Mary, two observers in motion relative to one another, make a fleeting contact. For example, they brush against each other as they pass. It clearly appears that Max and Mary are real with respect to each other. Then, suppose that event e_1 is intersected by Mary's simultaneity hyperplane; it follows that e_1 is real with respect to her. The transitivity of the relation 'x is real with respect to y' entails that, since e_1 is real with respect to Mary, and Mary is real with respect to Max, e_1 is also real with respect to Max. However, given the relativity of simultaneity, it could be that, although e_1 is simultaneous to Mary, it is located in the absolute future of Max. Indeed, since Max and Mary are in motion relative to one another, SR predicts that they will have different perspectives on whether some events, such as e_1, lie in the present or in the future. The situation is depicted in Fig. 4.5: the unprimed frame represents Max's perspective, and the primed frame (which is moving relative to the unprimed frame) represents Mary's perspective. Now, assuming that GBT is true, Max finds himself in an impossible situation: on the one hand, he

[26] An assumption is that what is present is comprised by simultaneous points/events. Nonetheless, there is a literature addressing the development of the Putnam-Rietdijk argument that precisely denies that, and identifies the present of an event with spacetime regions that contain light-like (or even spacelike) separated points.

[27] For the sake of accuracy, it was Rietdijk who actually formulated the argument in such a way that points in the future light-cone are dimmed to be real. In Putnam's version, you cannot get future light-cone points to be real, because I-now and you-now are taken to be located at the same spacetime point.

[28] However, the classical B-theorist is not totally immune to the Putnam-Rietdijk argument, since she believes in absolute simultaneity and absolute precedence (cf. Sect. 3.2).

Fig. 4.5 The Putnam-Rietdijk argument illustrated

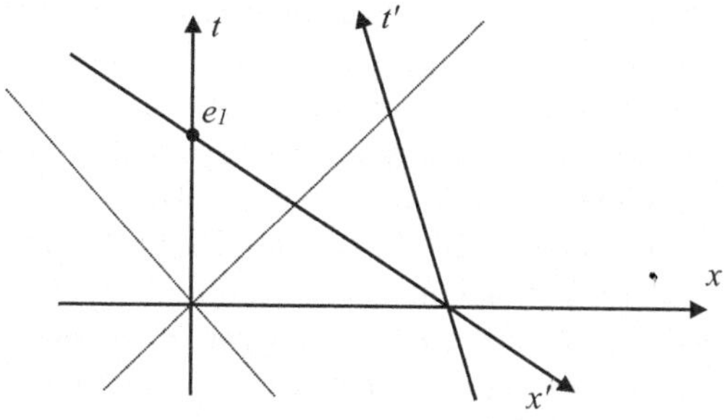

must take e_1 to be *unreal*, since it lies in his absolute future, but on the other hand, he must take e_1 to be *real*, since Mary takes it to be real and she is real with respect to him. This apparent contradiction leads Putnam to conclude that "[…] *future* things (or events) are already real" (1967: 242).[29]

The Putnam-Rietdijk argument seems to threaten the 2 essential components of GBT introduced in the previous chapter: (i) the asymmetry between the future and the past, and (ii) temporal becoming (cf. Sect. 3.4). In particular, since no frame of reference is privileged, it seems that all of the hyperplanes in the representation of the world are metaphysically on a par and, therefore, that there is no way to make sense of the thesis according to which there must be a geometric asymmetry between the future and the past (in order to ground some widespread intuitions about the nature of time) – the asymmetry collapses. More specifically, since no frame is privileged, no single hyperplane in the representation can be picked out from the set as being the unique 'present' axis around which reflection symmetry can be operated. The asymmetric-theories family, to which GBT belongs, seems therefore wrongheaded. Moreover, it seems that there is no way to make sense of temporal becoming, according to which new events are created in the present. After all, to say that temporal becoming is real is to say that "[…] there is something privileged about one of the many hyper-planes represented by the four-dimensional manifold" (Miller, 2013: 353), namely the latest hyperplane added to the growing block, which seems incompatible with SR – temporal becoming collapses. Mauro Dorato expresses the same thing in a slightly different way: "[t]o the extent that the notion of temporal becoming presupposes the unreality of future events as its necessary condition, [SR] seems to rule out also temporal becoming" (2008: 59).

However, although the Putnam-Rietdijk argument looks powerful, there are various options to avoid its conclusion. These options can be grouped into two families: the *compatibilist* and the *incompatibilist* options. Whereas the compatibilist options state that any metaphysical view of the world must be compatible with the fact that SR is approximately true, the incompatibilist options deny this. In the rest of the

[29] However, as Mauro Dorato rightly points out, a more accurate conclusion should be that "[…] past, present and future events all coexist *tenselessly*" (2008: 58).

section, I will review some of these options.[30] The aim is not be to identify one specific option as our best chance to defeat the Putnam-Rietdijk argument, but rather to show that the great variety of options strongly suggests that this argument is not definitive; there are plenty of ways that growing blockers can take to escape the pressure it exerts. In particular, I will discuss the *incompatibilist* options to be found in Zimmerman (2008), Bourne (2006), Miller (2013), and Tooley (1997), and the *compatibilist* options to be found in Sklar (1974), Stein (1968, 1991), and Correia and Rosenkranz (2018). At the end of the section, I will acknowledge that my preference goes to the *incompatibilist* family of options, because it seems better suited to accommodate both some of our intuitions about time, and some considerations from quantum mechanics (e.g., experimental results connected with John Bell's theorem). This preference will lead me to introduce, in the sixth section of the chapter, the causal set theory (CST), which, although still in its infancy, offers hope for reconciliation between GBT and our best physics.

First, let us focus on the *incompatibilist* options, which contest the metaphysical relevance of SR. These options come in 2 forms: (i) some philosophers argue that, although SR speaks to the geometry of spacetime, it has no ontological import, whereas (ii) some others go a step further by arguing that SR (at least as formulated above) is *not* approximately true of the world.[31] Dean Zimmerman (2008) choses an incompatibilist option of the first form. He argues that the four-dimensional manifold of spacetime points, which is posited by SR, is merely a theoretical entity and, therefore, should not lead to any ontological conclusion. For instance, the straightness of a worldline, he says, provides no information as to what exists; it merely indicates where an object would go (and where it could come from), if it were undergoing no accelerations (or decelerations). The fundamental information that the four-dimensional manifold gives is therefore not about ontology, but *accessibility*: it merely tells an object "[…] 'where to go next' if it is located at a series of points on the line, and no other forces are at work" (2008: 219). Interpreted in this way (rather than as a contribution to ontology), SR does no longer appear as a threat to GBT. An immediate objection could be that the geometry of spacetime seems fundamental to explain, e.g., the twin paradox (cf. Sect. 4.3); it is therefore not clear how Zimmerman can account for it.

In a similar vein, Craig Bourne (2006) and Kristie Miller (2013) argue that, although SR says that there is no privileged hyperplane, growing blockers are free

---

[30] These two families of options have been much discussed in the literature, cf. Bourne (2006) Christensen (1974), Clifton and Hogarth (1995), Correia and Rosenkranz (2018), Craig (2001a), Crisp (2003), Dainton (2010), Dorato (2006b), Godfrey-Smith (1979), Fine (2005), Hinchliff (2000), Markosian (2004), Mellor (1974), Miller (2013), Putnam (1967), Rakic (1997), Savitt (2000), Sider (2001), Sklar (1974), Stein (1968), Tooley (1997), Wüthrich (2010), and Zimmerman (2008).

[31] There is an obvious sense in which SR is not true of the world: "[…] once gravity is taken into account, SR must be replaced by GR which is arguably more fundamental […]" (Wüthrich, 2010: 265–266). However, this does not prevent SR from placing certain constraints on other theories (including theories of quantum gravity), such as, for example, the constraint that all physical possible interactions must be governed by Lorentz-covariant dynamics.

to reject the idea that this entails that "[…] all of the hyperplanes in our representation of the four-dimensional manifold are metaphysically on a par and hence that each corresponds to an *existing* hyperplane" (Miller, 2013: 353). Indeed, it is one thing to say that our best physics does not privilege any hyperplane, it is another to say that no hyperplane is *metaphysically* privileged. After all, the claim that there is a metaphysically privileged hyperplane (which seems both required by the geometric and the dynamic component of GBT) does not necessitate that we have (or could have) any physical or empirical access to *which* hyperplane is privileged. As Miller sums it up: "[…] what SR tells us is that it is in principle impossible to *determine* which plane is the metaphysically privileged one. But it does not tell us that no plane is in fact metaphysically privileged" (2013: 353). Growing blockers can therefore argue that, although SR entails that no hyperplane is *physically* privileged, there actually is a *metaphysically* privileged hyperplane that no physical or empirical method allows one to detect.

This first option sounds acceptable, but since our defense of GBT rests on our concern for accommodating a basic intuition we have regarding the nature of time (the future is open, while the past is fixed), it would surely be problematic to end up with a theory that says that, although there is a privileged hyperplane, it is empirically impossible to determine which one it is (cf. Prosser, 2000, 2007). As Miller herself notices: "[…] if there is no way to detect which plane is privileged and its being metaphysically privileged makes no empirical difference in the world, then it is hard to see how the fact that a plane is metaphysically privileged could ground our temporal phenomenology" (2013: 353). Indeed, how could it be that something that is empirically undetectable accounts for the manifest image of the world? It therefore seems that either the metaphysically privileged hyperplane is empirically salient, or GBT fails to account for our pre-theoretic thoughts about the nature of time. Thus, since one takes (at least part of) our temporal phenomenology seriously (i.e. as describing how the world *truly* is), and that this constitutes the core motivation for accepting GBT, it seems that one should reject Bourne-Miller's way of addressing the challenge raised by the Putnam-Rietdijk argument.

Instead, one could adopt an incompatibilist option of the second form and argue that, since there clearly seems to be a single universe-wide border between the past and the future, which coincides with the present, any theory that (frame-)relativizes it must be false (or incomplete). This seems corroborated by various considerations from quantum mechanics, especially by experimental results connected with John Bell's theorem, which suggest that some tenets of SR must be given up.[32] Without going into details, it indeed seems that if there are *truly* instantaneous connections between two correlated particles, A and B, at different places in space, then the objective notion of absolute simultaneity has a real application. For example, if B acquires determinate values (e.g., a determinate spin) by an independent

[32] John Bell's theorem proves the Einstein-Podolsky-Rosen thought-experiment (which aims to show that quantum mechanics is an 'incomplete' theory) to be wrong. In particular, Bell's theorem demonstrates that the hypothesis of local hidden variables is inconsistent with the way quantum systems behave (cf. Albert, 1994: 70).

measurement performed *simultaneously* on A (irrespective of the spatial separation of the particles), then, since this seems to require A and B to have a superluminal causal connection (sometimes called 'spooky-action-at-a-distance'), it can be concluded that SR, which explicitly rules out such connections, fails to provide a complete account of the spatiotemporal connections that actually exist.[33] In such circumstances, a natural belief is that SR will one day meet the same fate as Newtonian mechanics, i.e. being replaced by a theory with superior predictive and explanatory power. Such a theory will crucially diverge from SR in entailing that some events in our world do stand in relation of absolute simultaneity. Of course, this latter claim can be criticized, especially on the ground that relativistic empirical effects (e.g., 'length contractions' and 'time dilations') are well confirmed, and therefore that the successor theory will also have to account for them. The question is then whether this can be done with an objective notion of absolute simultaneity.

Some philosophers answer 'yes' to the latter question. For example, Michael Tooley sets himself the task of developing an alternative to SR, since "[…] the Special theory of Relativity does not provide a complete account of the spatiotemporal relations that obtain between events" (1997: 338). In his book *Time, Tense and Causation*, Tooley presents a theory that, although compatible with all the empirical data that are usually taken to confirm SR, entails that some events stand in relation of absolute simultaneity with respect to each other. This theory depicts an absolute substantival spacetime (i.e. a spacetime that is not reducible to spatiotemporal relations between events) that is causally self-propagating over time. As Tooley puts it: "[…] the fundamental argument for the central thesis that the world is a dynamic one in which the past and the present are real, but the future is not, rests upon a claim concerning the nature of causation – namely, that causation is a theoretical relation between events, and one whose basic postulates can only be satisfied in a dynamic world of the type in question" (1997: 376). Specifically, Tooley's theory is constructed from two well-established Einsteinian definitions:

(i) *The speed of light*, which is constant relative to absolute space,
(ii) *The relation of simultaneity*, according to which two events, e_1 and e_2, occur simultaneously relative to some frame of reference F iff light emanating from each object would arrive at an object O – which is both equidistant and at rest within F – *at the same time*.

But, whereas Einstein assumes that all inertial frames are on an equal footing and, therefore, that simultaneity is *relative*, Tooley denies this. According to him, if space is absolute, then some inertial frames should be at rest relative to it, so that one can define an objective notion of *absolute simultaneity*: "[t]wo events, [e_1] and [e_2], are *absolutely simultaneous* means the same as [e_1] and [e_2] are simultaneous relative to some frame of reference that is at rest with respect to absolute space" (1997: 343–344).

---

[33] Other considerations from quantum mechanics concern (i) the collapse of the wave-packet, which requires a privileged frame in which the collapse occurs and (ii) the Bohmian interpretation, which requires a privileged frame with respect to which non-local interactions are instantaneous.

Of course, the possibility of absolute simultaneity comes with consequences. The most important concerns the speed of light: whereas Einstein assumes that the numerical value of the speed of light is the same when measured in *all* directions in *all* frames of references (cf. the *Light Postulate*), Tooley notices that this can no longer be the case. In his theory, a light signal will have a speed of $(c - v)$ in a frame moving at velocity v in the same direction as the signal, whereas the signal will have a speed of $(c + v)$ in a frame moving at velocity v in the opposite direction to it (cf. 1997: 345–346). However, as Tooley insists, the rejection of the *Light Postulate* is not contradicted by empirical data. Actually, what empirical data support is a weaker postulate, the *Round-Trip Light Postulate*, according to which observers in all inertial frames will agree on the round-trip speed of a light signal travelling from any location L_1 to any other location L_2 *and back again*. By contrast, the *Light* Postulate rests, as Einstein (1905) himself concedes, on a mere convention, namely that the *one-way* speed of light (from L_1 to L_2) is constant in all inertial frames. Whereas many experiments have been undertaken to prove that *one-way* speed of light is constant, none has been successful yet; it is merely regarded as reasonable assumption.[34] Tooley feels therefore free to conclude that light might well travel at different speeds in inertial frames that are moving relative to absolute space.

This conclusion forces Tooley to explain *why* no such variations in the speed of light have ever been detected – recall the Michelson-Morley outcome. His answer relies on a Lorentz-style compensatory theory, according to which the variations in light speed are systematically concealed by the way natural processes are affected in moving frames. In short, since the original Lorentz transformations presuppose that the *one-way* speed of light is the same in all inertial frames, Tooley replaces them with new transformations, commonly referred to as '$\in$ – Lorentz transformations' (cf. Reichenbach, 1957: 127). Following John Winnie (1970), these new transformations are derived from a suitably modified version of SR (which has been shown to be inconsistent only if the original version of SR is itself inconsistent) that does not entail the *one-way Light Principle*. The $\in$ – Lorentz transformations allow Tooley to calculate the necessary compensation corresponding to different assumptions (captured by the variable '$\in$') concerning the relevant *one-way* speed of light (cf. Tooley, 1997: 349, and also Dainton, 2010: 337–342).

The details would bring us too far, but it is worth knowing that Tooley's neo-Lorentzian theory, enriched with further principles,[35] can account for all the same empirical effects as SR (including 'length contractions' and 'time dilations'). Now, although there is no definitive physical proof for there being an objective

[34] However, one may disagree with that. The literature on the conventionality of simultaneity (within any frame) notes that all such effort to measure one-way speed presuppose some standard of clock synchrony. Despite fringe controversy rumbling on, the overwhelming view is that there is no convention-free way of measuring one-way speed.

[35] Cf., in particular, 'The Principle of Absolute Fitzgerald-Lorentz Contraction' and 'The Principle of Absolute Time dilatation' (Tooley, 1997: 352–353).

relation of absolute simultaneity,[36] Tooley's theory might seem preferable to SR, since it is better suited to accommodate (i) a single universe-wide border between the past and the future (which seems reflected by our basic intuitions about time), and (ii) some considerations from quantum mechanics, especially experimental results connected with John Bell's theorem, which provide strong reasons to believe that "[…] there is sometimes no temporal gap between spatially separated events that are nomologically connected" (1997: 354).[37] However, despite these attractive features, Tooley's theory is often criticized for being both too *revisionist* and too *costly*, especially regarding its neo-Lorentzian nature. Christian Wüthrich, for instance, writes that "[Tooley's theory] violates Ockham's razor so crassly that the move cannot be justified by putting some post-verificationist philosophy of science on one's flag" (2010: 264). Ideally, the incompatibilist strategy would therefore be better served by a theory that retains the Lorentz invariance; this is precisely one of the qualities of the causal set theory, which will be explored in Sect. 4.6.

Second, let us focus on the compatibilist options, which take the metaphysical relevance of SR for granted. These options also come in 2 forms: (i) some philosophers challenge the premises of the Putnam-Rietdijk argument; (ii) some others reformulate GBT to make it expressible in relativistic settings. Lawrence Sklar (1974) choses a compatibilist option of the first form. He resists the claim that the non-physical relation '*x* is real with respect to *y*' is transitive. In that sense, from the fact that e_1 is real with respect to Mary (via the primed frame) and that Mary is real with respect to Max, it does not follow that e_1 is real with respect to Max (cf. Fig. 4.5). It indeed seems that, as far as the relation '*x* is real with respect to *y*' is intransitive, a growing blocker can acknowledge that different events exist relative to different frames, without being committed to the existence of events that are, from *her* frame of reference, located in the future. Sklar's move, which implies that what exists at a distance depends on a state of motion, seems thus to immediately block the Putnam-Rietdijk argument.

Of course, one might be reluctant to treat the relation '*x* is real with respect to *y*' as intransitive. One reason is that the rejection of transitivity comes with counterintuitive consequences. For example, "[…] two observers zooming past each other would share the same present without sharing what is real at a distance, and by simply changing reference frame (getting off a bus or jumping on an airplane), we would change what counts as real for us at a distance" (Dorato, 2008: 60). But one has to remember what SR taught us: all of our talk must be frame-relativized. Keeping this in mind, there seems to be no *a priori* reason to think that our talk about existence should be an exception – though Gödel famously claims that "[t]he concept of existence […] cannot be relativized without destroying its meaning

[36] Actually, the Lorentz symmetry seems to be well confirmed (cf. Will, 2005), which puts serious pressure on any approaches that requires a privileged frame of reference.

[37] As Tim Maudlin makes clear, "[e]mbedding quantum theory into the Minkowski space-time is not an impossible task, but […] the cost exacted by [the] theories which retain the Lorentz invariance is so high that one might rationally prefer to reject Relativity as the ultimate account of space-time structure" (2011: 201–202).

completely" (1949b: 558).[38] This reply may seem disappointing, but it must be recognized that one cannot have one's cake and eat it too: if one chooses the compatibilist path and, therefore, if one takes the metaphysical relevance of counterintuitive SR for granted, then there will be counterintuitive consequences, and the intransitivity of the relation '*x* is real with respect to *y*' might be one of them.

For those who regard the intransitivity of the relation '*x* is real with respect to *y*' as unacceptable but, at the same time, are inclined towards the compatibilist strategy, the situation is not altogether desperate. A compatibilist option of the second form has been developed by Howard Stein (1968, 1991), who aims to reconcile the unreality of future events with the idea that SR forces us to abandon the objective notion of absolute simultaneity. Basically, his idea is to show that events can be partitioned into past, present, and future, in a way that respects the geometric structure of Minkowski spacetime. To make that case, Stein establishes that, for any given point *x*, the only points that are real (or definite)[39] with respect to *x* are those that lie in *x*'s past light-cone. Conversely, *x*'s future light-cone and *x*'s absolute elsewhere contain points that are unreal (or indefinite) with respect to *x*.[40] In this context, 'being real (or definite)' is a *binary* relation between point-like events: $Rxy =_{df}$ '*y* is real (or definite) with respect to *x*'. This relation is *non-universal* (for all events *x* of spacetime, there are events *y* such that $\neg Rxy$), *reflexive*, and *transitive*. Then, Stein identifies the spatially extended present with the set of events on the past light-cone of the *here-now*. The latter proposition is motivated by an appreciation of epistemic accessibility: causal signals reaching us now emanate from the events on the past light-cone, and thus appear to us as being present. An immediate consequence of this option is the loss of a single universe-wide border between what is real (or definite) and what is not: two observers will not fully agree on what is real, unless their past light-cones fully coincide. There are therefore different perspectives for each observer, and what is real (or definite) for each of them is different.

Of course, Stein's theory provides no evidence that temporal becoming actually occurs in Minkowski spacetime, but it shows at least that, contrary to what the Putnam-Rietdijk argument suggests, the possibility of temporal becoming is not ruled out by SR. In particular, if one accepts to conceptually distinguish the notions of 'temporal becoming' and 'spatially extended present', so that the former does not necessarily requires the latter, SR no longer appears as a hostile environment for the creation of new things (e.g., spacetime points, events) in the present. Moreover, as suggested above, Stein's theory allows one to preserve the transitivity of the relation '*x* is real with respect to *y*'. Indeed, given that a point can only be real with respect to points that are in its causal past, although it follows from *Rab* and *Rbc* that *Rac*

---

[38] Kit Fine (2005: §10) also suggests to relativize existence when he argues in favor of a frame-theoretic form of non-standard realism. According to his view (fragmentalism, cf. Sect. 3.2), what is present is relative to a frame of reference and, therefore, what exists is fragmented in that it depends on the choice of frame (recovered as a maximally coherent collection of facts).

[39] Following Maxwell (1985: 24), by "definite", Stein means "ontologically fixed".

[40] Stein assumes that Minkowski spacetime allows for a definite time-orientation, which is needed to make sense of the claim that what is future (rather than past) is not yet definite.

(transitivity), there is no risk of 'spreading reality' from the past to the future, since *c* will also be in the absolute past with respect to *a*. By contrast, the relation '*x* is real with respect to *y*' is not symmetric: if *b* is real with respect to *a* (because *b* lies in *a*'s causal past), it does not follow that *a* is real with respect to *b*; otherwise, we would end up in the situation where all points are equally real with respect to each other. This is no big deal, however, since every version of GBT (including Broad and Tooley's versions) must anyway take the relation '*x* is real with respect to *y*' to be non-symmetric: it is a milestone of GBT that "[…] 1900 is real as of 2000, but as of 1900, the year 2000 is not real" (Dainton, 2010: 333).

At first sight, Stein's theory appears to be coherent, but it comes with important costs. First, although Dorato (1995, 2008), Dainton (2010), and Correia and Rosenkranz (2018) provide some suggestive descriptions, it might be difficult to visualize what this theory amounts to; intuitive pictorial flesh can hardly be put on it, or so it could be argued.[41] Second, as Christian Wüthrich and Craig Callender point out, Stein's theory has deeply counterintuitive consequences. For instance, it entails that "[…] if we take the past lightcone as the present, then the big bang counts […] as 'now'" (2016: 5). However, the latter objection can be avoided simply by denying that the present is identical to the set of events in the past light-cone of the *here-now*. Indeed, following Dorato (2008) and Correia and Rosenkranz (2018), Stein's theory is more charitably interpreted as a pointy relativistic version of GBT. Third, following Savitt (2000: 568), Correia and Rosenkranz (2018: 151–152) argue that this pointy relativistic version of GBT leaves no room for the existence of spatiotemporally-extended things that one can perceive. They provide the following example. I (understood as an embodied consciousness occupying a single spacetime point) can *here-now* perceive my limbs to be located at distinct spacetime points. Obviously, since the light signals involved in this perception take time to reach me, what I actually perceive is that, somewhere in the immediate causal past, my limbs occupy distinct spacetime points. This might not appear problematic since, unlike *here-now*, the causal past is thick enough to comprise several spacetime points, but the situation is quite different when *persistence* comes into the picture. Since I have every reason to believe that my limbs have persisted and, therefore, that they exist *here-now*, the question that arises is: 'Where are my limbs located?'. No convincing answer is available to Stein. Whereas I *here-now* perceive my limbs to have been located at distinct points in the causal past of *here-now*, Stein's theory provides no plurality of points in the causal future of those points that my limbs could occupy *here-now*.

Finally, a third compatibilist option has been developed by Fabrice Correia and Sven Rosenkranz (2018). In their book *Nothing to come*, they reformulate GBT without using the problematic notion of presentness, so the theory can

[41] It is true that Stein remains evasive when it comes to describing the metaphysical picture of the world that his theory involves. He merely talks of "[…] a notion of temporal evolution as (in some sense) *a becoming real, or becoming determinate, of what is not yet real or determinate*" (1968: 14), and of a distinction between a part of the world history that "is ontologically fixed" and a part that "is not yet settled" (1991: 148). However, a sketch of Stein's model can be found in Fig. 4.6.

accommodate relativistic spacetime structure. To make that case, Correia and Rosenkranz define GBT as describing a world in constant creation in which nothing ever gets annihilated: while new things continuously come into existence, nothing ever goes out of existence. This results from the acceptance of 2 principles that we have already encountered in Sect. 3.3:

(P1) Always everything will always in the future be something (E!*x* →GE!*x*)
(P2) Every time is new at itself' (T*x* → At *x*, H¬E!*x*).[42]

This reformulation brings many benefits. First, (P1) and (P2) are apparently sufficient to capture the idea of a single edge of becoming beyond which nothing exists (cf. 2018: 44–45). Second, since neither (P1) nor (P2) invokes a notion of 'being present', this version of GBT avoids the main pitfalls encountered by traditional attempts to spell out this notion (cf. Sect. 3.3). Third, when physics come into the picture, this version of GBT seems well suited to meet the relativistic requirements. In particular, since no objective notion of absolute simultaneity is required to express (P1) and (P2), it dissipates the tension highlighted by the Putnam-Rietdijk argument. However, this cannot be the end of the story: the process of 'constant creation', involved in this version of GBT, rests on an absolute temporal order relation, which seems inconsistent with relativistic conceptions of spacetime. That is why Correia and Rosenkranz propose to characterize GBT in a spacetime-sensitive language definable in relativistic terms. Roughly, this means that (P1) and (P2) should be replaced with the following two relativistic principles:

($P1_R$) Everywhere everything everywhere in the causal future still exists (E!*x* → ▼E!*x*, where '▼φ' stands for 'Everywhere in the causal future, φ').
($P2_R$) For any spacetime point *s*, at *s*, everywhere in the causal past of *s*, *s* did not yet exist' (S*x* → @ *x*▲¬E!*x*, where 'S' abbreviates the predicate 'is a spacetime-point', and '@ *m*' abbreviates 'At spacetime-point *m*').

Specifically, ($P1_R$) and ($P2_R$) imply that, for any spacetime point *s*, there are spacetime points in the causal past of *s*, but no spacetime points in its causal future (cf. 2018: 149). As a last step, Correia and Rosenkranz add to ($P1_R$) and ($P2_R$) the principle (BO) 'for any spacetime point *s*, at *s*, everywhere in the elsewhere region of *s*, *s* exists' (S*x* → @ *x*◀E!*x*, where '◀' stands for 'Everywhere in the elsewhere region'), so they can derive that, for any spacetime point *s*, the elsewhere region of *s* may be populated.[43] This brings a further benefit: unlike Stein's theory, this theory can account for the persistence of non-causally separated things that we can perceive. In particular, it allows for the existence, *here-now*, of several spacetime points that my limbs, which I perceive to have been located at distinct points in the causal

[42] As Correia and Rosenkranz make clear, (P2) "[…] guarantees that any newly added resident of time is located at the time of its addition" (2018: 46).

[43] Interestingly, when one replaces (BO) by (PO) 'for any spacetime point *s*, at *s*, everywhere in the elsewhere region of *s*, *s* does not exist' (S*x* → @ *x*◀¬E!*x*), one obtains a theory close to Stein's pointy relativistic version of GBT, introduced above (cf. Correia & Rosenkranz, 2018: 149).

past, can occupy. The similarities and differences between Stein's relativistic version of GBT and that of Correia and Rosenkranz are pictured in Fig. 4.6.

However, here again, one ends up with a theory, so-called 'Bow-tie relativistic GBT', which seems immune to the usual objections drawn from relativistic physics, but which requires us to abandon some of our basic intuitions about time. This is what prompts Ulrich Meyer, for instance, to say that "[…] I am not sure that [Correia and Rosenkranz's] proposal is what other people had in mind when they endorsed the growing block theory" (2019). For example, since it allows for two points, which are causally separated from *here-now*, to be in a relation of causal precedence, Bow-tie relativistic GBT entails that the relation '*x* is real with respect to *y*' is intransitive. This clearly betrays some of our common intuitions, especially the intuition that "[i]f it is the case that all and only the things that stand in a certain relation *R* to me-now are real, and you-now are also real, then it is also the case that all and only the things that stand in the relation *R* to you-now are real" (Putnam, 1967: 241). Although Correia and Rosenkranz contest the appeal to common intuitions in those circumstances,[44] one might be reluctant to pay such a price for the reconciliation of GBT with our best physics. For, to sacrifice our intuitions is to sacrifice what made GBT attractive in the first place.

At this step, the idea is not to defend one option (either compatibilist or incompatibilist) as the best way to meet the challenge raised by the Putnam-Rietdijk argument. Rather, the idea is to show that, although the objection to GBT based on relativity is troubling – mainly because the progress of science has taught us to be extremely wary about the deliverances of our intuitions – there are various reasons to think that SR does not extinguish all hope of adequately describing reality as growing. Both incompatibilist and compatibilist options are available to defuse the tension between SR and GBT; they are summarized in Fig. 4.7. Nonetheless, I am inclined to think that incompatibilist options are more promising than compatibilist ones, because, as far as our intuitions are concerned, accepting SR (at least as

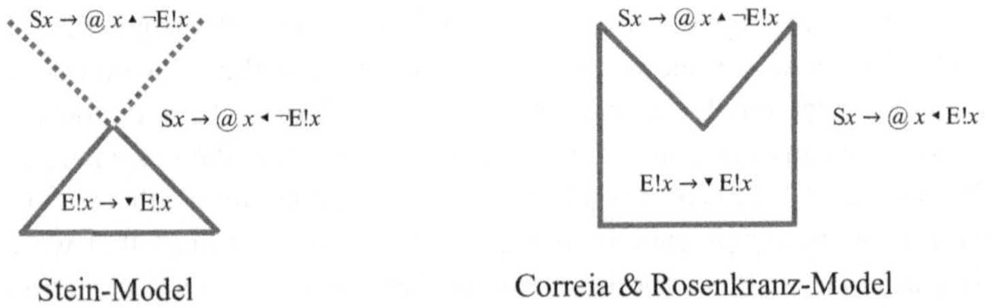

Fig. 4.6 Stein's pointy relativistic model, and Correia and Rosenkranz's bow-tie relativistic model. (This figure is freely inspired from Correia & Rosenkranz's figures 9.3. and 9.4 (cf. 2018: 150))

[44] After all, as Correia and Rosenkranz put it: "[…] the idea that there is a region of points non-causally separated from here-now some of which stand in relations of causal precedence to one another is itself not very intuitive by everyday standards" (2018: 156).

Incompatibilist options	Compatibilist options
Zimmerman (2008): SR is about *accessibility* and should therefore not lead to any ontological conclusion.	**Sklar** (1974): The relation 'x is real with respect to y' is *intransitive*, and so the Putnam-Rietdijk argument must be rejected.
Bourne (2006), **Miller** (2013): SR entails that no hyperplane is *physically* privileged, while this is compatible with there being a *metaphysically* privileged hyperplane.	**Stein** (1968, 1991): What is real as of a given time and place (or spacetime point) consists of the events at that time and place, together with the events in the past light-cone.
Tooley (1997): SR should be replaced with a neo-Lorentzian theory that posits a real relationship of absolute simultaneity.	**Correia & Rosenkranz** (2018): GBT should be reformulated in a spacetime-sensitive language (with no reference to the notion of 'being present'), such that, for any spacetime point s, there are spacetime points in the causal past and in the elsewhere region of s, but not in its causal future.

Fig. 4.7 Incompatibilist and compatibilist options summarized

formulated above) requires a tremendous amount of sacrifices. Does the previous statement undermine any attempt to reconcile the manifest image with our best physics? The answer is 'not necessarily', since it is far from clear that SR belongs to our best physics. As has been said, many considerations from quantum mechanics (e.g., experimental results connected with John Bell's theorem) lead us to think that SR should be replaced by a theory with superior predictive and explanatory power. A reply might be that SR is anyway not the final answer that Einstein provides to the question of the nature of spacetime. But, as will be shown in the next section, the General theory of relativity (GR), which has been proved incomplete anyway,[45] also encounters important difficulties when squaring it with quantum mechanics. That is (partly) why we will turn our attention to the nascent theories of quantum gravity in Sect. 4.6, and especially to the causal set theory (CST). As will become clear, this final move should be regarded as a representative of the incompatibilist options.[46]

4.5 Beyond Special Relativity

In the brief history of modern physics that has been told so far, two successive phases were identified. The first phase was (neo-)Newtonian mechanics, which, since it takes B-relations like '*earlier than, later than* or *simultaneous with*' to be

[45] As will be made clear, within GR, one can prove theorems that show that, under very general conditions, singularities (i.e. borders of spacetime beyond which geodesics cannot be extended) are unavoidable (cf. Hawking & Penrose, 1996; Kiefer, 2011).

[46] A similar dialectic can be found in Monton (2006), but whereas Monton is mainly interested in so-called fixed-foliation approaches to quantum gravity (in order to claim that fundamental physics is hospitable to presentism), I will argue that (at least a form of) temporal becoming could be restored by a dynamics, the so-called 'classical sequential growth dynamic' (CSG), by which the growth in the causal sets takes place.

absolute (or frame-invariant), allows for an objective and absolute temporal ordering of events. The second phase was Special Relativity (SR), which, since it admits no objective notion of absolute simultaneity, comes with a loss of comparability: for two space-like separated events, e_1 and e_2, there no longer is a frame-independent fact of the matter as to whether e_1 is, for instance, *earlier* than e_2. Accordingly, the temporal ordering of events is only partial. In this section, one will introduce a third phase, General Relativity (GR), which makes things even more radical: since the topology of GR allows for the possibility of closed time-like curves, it entails that the temporal ordering of events is not even partial. As Christian Wüthrich puts it: "[i]n fact, there is no global time deserving this title in general relativity (GR), a fact that finds a particularly vivid expression in the so-called 'problem of time' arising in the Hamiltonian formulation of GR" (2010: 260). There therefore is no reason to believe that GR may help the compatibilists, who rightly point out that counterintuitive SR is not the final answer that Einstein provides to the question of the nature of spacetime, to restore the manifest image. Worse, since GR also requires '*relativistic causality*' (i.e. that causal signals can travel at most as fast as light), it seems to be in no better position than SR when squaring it with quantum mechanics. As a reminder, Bell's theorem predicts that quantum mechanics is non-local, in the sense that a measurement on a system by an observer at one location has an instantaneous effect on a distant correlated system (one with which the original system has interacted).

But let's start with some basics. Ten years of hard work after the development of SR, Einstein completed a new theory, General Relativity (GR), which is not to be understood as a disavowal of his early work, but rather as an attempt to overcome its limited aspirations. Whereas SR provides a rigorous account of the difference between accelerated and inertial motions (reflected by Minkowski spacetime's 'affine' structure) and of the dynamics of moving objects, it remains silent on one crucial phenomenon: gravitation. The reason is roughly that Newton's theory of gravitation, which was then widely accepted, could hardly be slotted into SR's framework, because it is incompatible with (i) some relativistic effects (such as 'length contractions' and 'time dilations'), and (ii) the *Light Postulate*. This is easily understandable. The strength of Newtonian gravitational force between two objects depends on their distance apart, while SR states that this distance will be different in the inertial frames of the relevant objects (assuming that they are moving relative to one another). Therefore, two observers on the two objects will obtain different results when measuring the strength of the gravitational attraction between them. Moreover, Newtonian gravitation acts *instantaneously* between all material objects, irrespective of their spatial separation, while this is incompatible with SR's limited causal propagation – recall the light-cone structure of Minkowski spacetime (cf. Sect. 4.3). Behind the apparent incompatibility between Newton's theory of gravitation and relativity lies a fundamental disagreement on the nature of gravitation: unlike Newton, Einstein does not regard it as a *force* (that allows for instantaneous 'action-at-a-distance'). Instead, Einstein will propose a new conception of gravitation, according to which gravitation is a mass-induced spacetime curvature, that is at the core of GR.

Basically, GR rests on the extension of *the Relativity Postulate* from inertial motion to accelerated motion (or so Einstein thought): the laws of nature do not distinguish between frames that are freely falling in gravitational fields and inertial frames. The reason is that "[…] the inertia of a test mass is increased if it is surrounded by a shell of inertial masses and that, if these same masses are accelerated, they tend to drag the test mass with it" (Norton, 1993: 799). For example, if Mary is in a windowless spaceship (enjoying weightless conditions), there is nothing she can do to determine her state of motion: perhaps she is accelerating towards a nearby black hole, perhaps she is drifting in space. In either case, all the experiments she could perform within her spaceship will yield exactly the same results. This is known as *the Equivalence Principle*, which was originally expressed as follows:

> […] we assume that the systems K [inertial system in a homogeneous gravitational field] and K' [uniformly accelerated system in gravitation free space] are physically exactly equivalent, that is, […] we assume that we may just as well regard the system K a s being in a space free from gravitational fields, if we then regard K as uniformly accelerated (Einstein, 1911, §1).[47]

A significant consequence of this principle (which is also true in Newton's theory) is that it makes distinguishing between gravitational and acceleration effects impossible. For example, if Mary realizes that she is no longer weightless (and that objects around her are no longer floating in mid-air, but are lying on the floor), no experiment she could carry out in her windowless spaceship would help her to determine whether this be a consequence of gravity (e.g., her spaceship is motionless near a planet) or acceleration (e.g., her spaceship is accelerated by a rocket). This led Einstein to infer that gravity and acceleration might essentially be the same phenomenon: a modification of the very structure of spacetime.

Another consequence of *the Equivalence Principle* directly impacts light: although light is known to be massless, the paths it takes are deflected by gravity. In particular, given that a light signal travelling in an accelerating spaceship will follow a curved path (e.g., it will hit the opposing wall at a lower position than if the spaceship were not moving), and that gravity has the same (local) effects as acceleration, a light signal travelling in a stationary spaceship within a gravitational field will follow a similar trajectory (e.g., it will also hit the opposing wall lower down). This observation has enormous consequences. As a reminder, Minkowski spacetime is built around the paths that light can take (cf. Sect. 4.3). In that sense, these paths do not only define the absolute future and past of every point, but they are also the paths of the shortest possible distance: "[…] the interval between light-like connected points is zero (and so there is, from the perspective of light, no spatial or temporal distance between them at all)" (Dainton, 2010: 346). Therefore, supposing that (i) Minkowski spacetime is approximately true of the world, and (ii) gravity has the same (local) effects as acceleration, it can be inferred that gravity affects the structure of spacetime itself. Actually, Einstein even goes a step further by arguing that gravity is *nothing but* the warping of spacetime: the effects that Newton

[47] The translation is to be found in Lorentz et al. (1952: VI, §1).

explained in terms of an attractive force operating over material objects are to be understood as the result of matter bending spacetime in its vicinity – the more matter, the greater the distortion. And this matter-induced distortion is not transmitted through spacetime instantaneously, but at the speed of light; Einstein thus gets rid of action-at-a-distance.

Specifically, a large concentration of matter induces a strong curvature of spacetime in its immediate vicinity, and the curvature is transmitted through spacetime from region to region (which are not directly affected by the matter) in a gradually weaker way. As Dainton puts it: "[a]s you get further away from the mass, the curvature of each successive spherical region of spacetime diminished" (2010: 347). The overall shape of spacetime is therefore the product of the combined influence of all the material objects, and this shape evolves as the objects move. The following example might help to clarify the situation. As has been seen, the natural path of an object, which is not subject to any force, is a straight line (cf. Sects. 4.2 and 4.3). A spaceship moving along its natural path will therefore continue to do so (without slowing or speeding up) until some external force acts on it. Now, suppose that this spaceship quickly flies past a planet; its trajectory will be deflected. The planet will cause the spaceship to move in a curve rather than a straight line. Yet, GR states that the spaceship does not experience any external force (assuming that it is not accelerated by its engines). How is that possible? GR's answer is this: the planet's mass alters the natural path of the spaceship. Counterintuitively, the planet does not pull the spaceship toward it (*pace* Newton), but affects the geodesics[48] of spacetime (which deviate from Euclidean straight lines), and hence modifies its shape. As a consequence, gravity produces not only space-bending effects, but also time-dilation effects: "[…] clocks tick more slowly in the vicinity of large material objects; the stronger the spacetime curvature, the slower the clocks tick" (Dainton, 2010: 350).

This could make GR's spacetime look very different from both Newton and Minkowski's spacetimes, which are entirely unaffected by the presence and distribution of matter within them. But it is worth noting that, in GR, the metric can always be approximated to the Minkowski form, at least in small regions of spacetime. However, on the large scale, gravity can no longer be neglected and, therefore, the manifold should not be expected to be flat. The light-cones must be bent toward the location of mass, with greater curvature near larger masses. As will become clear, many different topologies are consistent with what equations of GR (or Einstein field equations, thereafter) tell us about matter-distributions and spacetime curvatures, and some of them are really bizarre. For instance, GR does not rule out the possibility of 'closed time-like curves' (i.e. paths through spacetime that loop back upon themselves), although such non-orientable scenarios are often regarded as "not physically real" or "pathological" (but not "unphysical") (cf. Zimmerman, 2011: 188).

But first things first: what is a spacetime within GR? A spacetime is a four-dimensional manifold with a Lorentzian geometry. The Lorentzian geometry gives

[48] Geodesics are the shortest paths between two points in a Riemannian manifold.

spacetime its light-cone structure. Assuming that spacetime is globally hyperbolic (which corresponds to a certain condition on the causal structure of a variably curved spacetime), the causal interpretation of light-cones is still valid in GR: the future light-cone of a certain event includes the boundary of its causal future (and similarly for the past). But, given gravitational lensing, which occurs when a huge amount of matter (e.g., a cluster of galaxies) creates a gravitational field that distorts light, it might be that the light-cone folds on itself. The various ways in which spacetime can bend and twist are captured by complex equations: Einstein's field equations. They were first formulated in the form of a tensor equation, which relates the curvature of spacetime (described by the Einstein tensor) with local energy, momentum, and stress, within that spacetime (described by the stress-energy tensor). In other words, Einstein's field equations establish a systematic relationship between the geometry of spacetime and the distribution of mass-energy through spacetime. They thus determine the metric tensor of spacetime (i.e. all the geometric and causal structure) for a given arrangement of stress-energy-momentum in the spacetime. The solutions of Einstein's field equations are the components of the metric tensor. It is worth noting that, since the Einstein tensor has ten independent components, the Einstein field equations can be written as a set of ten non-linear partial differential equations; this will be useful in Sect. 4.6 to understand how causal set theorists retrieve the metric, the topology, and the differential structure of the manifold from the causal structure and volume information.

One of the basic principles of GR is its general covariance: its laws remain unchanged under an arbitrary transformation of the spacetime coordinates. This means that, in GR, "[…] coordinates on spacetime have no physical significance, no more significance than a choice of coordinate grid on a map of Mexico City, say" (Dowker, 2020: 145). This is, in particular, the aversion to violation of general covariance, which leads most philosophers of physics to claim that GR forces a 'block universe' view on us. It must indeed be acknowledged that GBT, for instance, contradicts general covariance by providing a foliation of spacetime into space-like hypersurfaces: "[i]t picks out a special time coordinate labelling the leaves of the foliation and gives it physical significance" (Dowker, 2020: 145). In that respect, GR seems just as at odds with our intuitions as its predecessor. After all, whether it is in GR or SR, simultaneity (when defined by Einstein's light-signaling method) is relative, although this relativity is expressed differently.[49] Therefore, a hypersurface of becoming, as depicted by traditional versions of GBT (to account for some of our intuitions), is equally problematic in both theories.[50] Worse, since GR allows for the

[49] Whereas in SR the relativity of simultaneity is expressed through the notion of inertial frames (cf. Sect. 4.3), in GR there are in general no exact and global inertial frames anymore. A reference frame will thus have to satisfy weaker conditions to qualify as acceptable. Although there are various methods for doing this, all of them generally provide multiple equally physical frames none of which can be privileged. Any two events that are simultaneous with respect to one such frame will generally not be simultaneous in any other frame.

[50] However, it seems that a certain class of GR models (viz. the Friedmann-Robertson-Walker models), which assume an even distribution of mass-energy throughout the universe, allows one to

possibility of 'closed time-like curves' (and hence entails that the temporal ordering of events is not even partial), the situation even appears more unfavorable for GBT than it was in SR. For example, Kurt Gödel (1949a) argues that the mere fact that closed time-like curves are possible suffices to establish the 'block universe' view of time: if closed time-like curves can be produced in our universe by rearranging matter,[51] then the past and future times to which these curves would provide access *must be real*. This does not mean that there is no version of GBT that can accommodate relativistic spacetime (although most versions of GBT take the causal precedence relation to be irreflexive, and hence exclude closed causal curves),[52] but this means that these versions of GBT must make so many concessions to relativity that they inevitably lose the 'intuitiveness' that made GBT attractive in the first place.

Thus, it seems that the compatibilist strategy – which rests on the idea that the theory of relativity is (approximately) true of the world and, therefore, that any metaphysical view of the world must be compatible with it – condemns us to sacrifice our most basic intuitions on the nature of time. Taking that for granted, it seems that the only real hope to preserve our intuitions from the pressure exerted by physics is to opt for an incompatibilist strategy: although GR is undoubtedly closer to the truth of the world than SR (it accommodates the gravitational effects, after all), it fails to provide an accurate description of it. But is there any naturalistic reason to privilege such an incompatibilist strategy? The answer seems to be 'yes'. Given that the causal interpretation of light-cones is still valid in GR, causal processes or signals can travel at most as fast as light, while this requirement, sometimes referred to as '*relativistic causality*', can apparently be violated in the quantum context. Indeed, the interpretative problems of any quantum theory about non-locality and the measurement problem[53] are widely regarded as threatening relativistic causality. As, for instance, Jeremy Butterfield puts it: "[…] non-locality looks like 'spooky

assign a global (or cosmic) time to all events, despite the fact that the relevant events are space-like separated. The manifolds in the models in question can be "[…] exhaustively partitioned into foliations of non-intersecting global (three-dimensional) hyperplanes (or 'leaves') that are orthogonal to the time-like geodesics" (Dainton, 2010: 381). Therefore, assuming that all the points on each of these hyperplanes are simultaneous, this distinctive foliation seems to generate a universe-wide temporal ordering for events. This leads some philosophers, e.g., John Lucas (1999), to defend a version of GBT, which depicts the universe as a solid sphere, growing as hyperplane is added to hyperplane.

[51] For instance, leaving aside the formidable technological difficulties, Kip Thorne (1994: ch. 14) argues for the possibility of *wormhole* time-machines, which seem to provide an *empirical* way to determine whether the past and the future exist: these *wormhole* time-machines only work in a block universe.

[52] This applies to 'classical' GBT: the temporal precedence relation cannot define a circular temporal order if GBT is true. The reason is roughly that moving forward in the 'future' direction inflates the ontology. As a result, one cannot, by moving forward, reach a spacetime point one previously was.

[53] The 'measurement problem' refers to the lack of a precise principle to decide which evolution of a quantum system will arise. The best-known example is 'Schrödinger's cat' (cf. Maudlin, 2019: 97–98).

action-at-a-distance'; and if measurement involves a 'collapse of the wave-packet' perhaps the collapse is superluminal'" (2007: 302, cf. also Maudlin, 2011: 187).

Specifically, since relativistic spacetime typically has no preferred foliation or 'slicing', it conflicts with at least 2 considerations from quantum mechanics. First, the correlation (or entanglement) of distant systems, which is widely accepted as a quantifiable and exploitable physical resource, produces a violation of Bell's inequality in quantum theory. In that sense, whereas Erwin Schrödinger (in agreement with Albert Einstein) claims that "[m]easurements on separated systems cannot directly influence each other – that would be magic" (1935: 161), John Bell proved that the magic is real: although GR prohibits anything from traveling faster than light, it must be recognized that non-locality is inherent in the quantum theory; a measurement on a system at one location has an instantaneous effect on a distant correlated system. Second, a collapse of the wave-packet, which refers to an irreducible change in the state of an isolated quantum system (contravening the deterministic and continuous evolution prescribed by the Schrödinger equation),[54] makes use of absolute simultaneity in specifying the collapses of the quantum state. For example, as Maudlin (2019) makes clear, the dynamics of GRW (the most prominent collapse theory) employs the notion of absolute simultaneity in specifying the collapse dynamics.[55] Admittedly, the collapse of the wave-packet represents just one of the several families of interpretations of quantum theory, but it is not clear that the other families of interpretations are in a better position. For instance, Bohmian mechanics, which adds to quantum theory's deterministic evolution of the wavefunction the postulate that certain preferred quantities have at all times a definite value, presupposes an absolute time structure: it "[…] makes essential use of the objective time order of distant events" (Maudlin, 2019: 205). These considerations strongly suggest that GR, which explicitly rules out the possibility of absolute simultaneity, is not the final word on the nature of spacetime. Interestingly, the issue of non-locality and that of measurement are interrelated: non-locality appears when one solves the measurement problem. For example, "[i]f one resolves the measurement problem by allowing a real physical process of wave collapse, it is the collapse dynamics which manifests the non-locality […]. If one resolves the measurement problem by postulating additional variables beside the wave function, it is the dynamics of these variables which manifests the non-locality […]" (Maudlin, 2011: xi).

Before we go any further, it is important to recall the current state of knowledge in physics. SR is, at best, only approximately true. For more than one century now, GR has looked like our best theory of the structure of spacetime. Progressively, the difficulties in squaring GR with quantum mechanics have become salient to more

[54] 'The collapse of the wave-packet' covers, among other items, the Copenhagen interpretation (whatever that is) and the program developed by Giancarlo Ghirardi, Alberto Rimini, and Tulio Weber, which is known as GRW.

[55] However, Roderich Tumulka has recently developed a fully relativistic version of GRW that can produce violations of Bell's inequality for experiments at space-like separation (cf. Maudlin, 2011: ch. 10).

and more researchers working on the fundamental structure of reality. An idea might therefore be that our most fundamental physics is rather to be found in the nascent theories of quantum gravity. This idea serves the incompatibilist strategy: perhaps, as the Putnam-Rietdijk argument points out, GBT (or more accurately the unreality of the future) is at odds with Einstein's theories, but the fact that GR is probably not the final word on spacetime structure suggests that a more relevant question to ask is whether GBT is compatible with the development of a quantum theory of gravity. After all, it might be that both SR and GR are hostile to (at least intuitive versions of) GBT, but that a specific quantum theory of gravity is hospitable to it. To put it another way, even if GBT turns out to be incompatible with both SR and GR (which again is doubtful), it in no way follows that GBT is incompatible with our most fundamental physics. In this respect, the next section will be devoted to studying of the causal sets theory, which promises to offer (through a still-classical dynamics) a naturalistic basis to GBT (without betraying the principle of general covariance).

4.6 Quantum Gravity and the Revival of Temporal Becoming

According to many theoretical physicists, one of the yet outstanding tasks in fundamental physics is the development of a quantum theory of gravity. The so far unsuccessful attempt to develop such a theory is an attempt to unify the General theory of relativity (GR) with the principles of quantum mechanics (QM). In brief, Quantum Gravity aims to describe everything in the universe in terms of Quantum Theory.[56] Although the quest for unification is often presented as the main motivation behind the search for Quantum Gravity, it is worth noting that it is not the only motivation, and perhaps not even a good motivation (since it rests on a purely inductive basis).[57] As many physicists insist, the most compelling reason *why* one wants to pursue a quantum theory of gravity is that there are phenomena in which both gravitational and quantum effects should play an irreducible role. For example, it seems that the very early universe and the dynamics of black holes, which both combine high energy densities and strong gravitational fields, cannot be understood without a theory that coherently models the interaction of quantum matter with strong gravitational fields.

Although it should be acknowledged that the search for a quantum theory of gravity is currently dominated by two research programs, String Theory and Loop

[56] This is not equivalent to 'quantizing gravity' (although quantizing gravity is one way to obtain a quantum theory of gravity), see Wüthrich, 2005.

[57] Specifically, the motivation of unification (all physics, especially all the forces, have to be treated in the same theoretical unified framework) merely rests on physical past successes (e.g., Maxwell demonstrated that the electric and magnetic forces are aspects of a single electromagnetic force) and, therefore, should not be seen as an *a priori* truth (but at best as an *a posteriori* fact). Moreover, gravity could turn out not to be a force.

Quantum Gravity, a third program, called 'Causal Set Theory' (CST), arouses increasing interest, especially among those who aim to rescue the manifest image from relativity (cf. Dowker et al., 2004, Sorkin, 2007).[58] These various research programs reflect divergences of opinion regarding the nature of time; but it seems that these divergences have diminished in the last few years, and that many conclusions have become reasonably clear to most. According to Carlo Rovelli, for example, "[w]hat has been clarified is that the residual temporal scaffolding of general relativity [...] falls away if we take quanta into account" (2018: 94). In that sense, there seems to be a minimum scale for all phenomena, including time. The value of minimum time, called 'Planck time', is estimated at 10^{-44} seconds. This implies that if we could measure the duration of an interval with the most precise clock imaginable, we should find that the time measured takes only certain discrete, special values. As Rovelli says, time should therefore not be thought as "[...] something that flows uniformly but as something that in a certain sense jumps, kangaroo-like, from one value to another" (2018: 96). This is comprehensible: given that granularity is ubiquitous in nature (e.g., light is made of photons, the electrons in atoms can only take on certain discrete values of energy, etc.), it seems natural to suppose that space and time are granular too – though there are also many continuous quantities (at least in quantum mechanics), e.g., position, momentum and the like.[59]

The idea that space and time are fundamentally discrete is not new. It goes back at least to Zeno's paradoxes (fifth century BC), which aim to show that the limitless divisibility of space and time leads to a contradiction and, therefore, that the apparently evolving reality should rather be conceived as a static, changeless unity. In reply, Leucippus and Democritus argued that the successive division of space and time is *not* limitless, but terminates in atoms (understood as particles incapable of being further divided) and, therefore, that the continuous is reducible to the discrete (cf. Bell, 2013: §1). In the seventeenth century, Leibniz famously held that space and time, as continua, are ideal, and that anything real, in particular matter, is discrete, compounded of simple unit substances called 'monads'. In the 19th century, Bernhard Riemann listed some benefits of taking the deep structure of space to be discrete rather than continuous. In particular, he argued that whereas counting the elements composing a region of discrete space provides a natural measure of that region's volume, a continuous space lacks this possibility and therefore requires that the origin of the metric relationship be explained in some other way. Finally, the subsequent development of physics has provided compelling reasons for questioning the continuum, including the singularities and infinities of GR, QM and black hole thermodynamics. Einstein was, by the way, one of the first to voice doubts of this sort. As he put it: "[i]f the molecular view of matter is the correct (appropriate) one, i.e., if a part of the universe is to be represented by a finite number of moving

[58] It must be acknowledged that neither String Theory, nor Loops Quantum Gravity seems to be compatible with temporal becoming (cf. Callender, 2000; Wüthrich, 2010).

[59] This widely alleged view has nonetheless been criticized, see for example Esfeld (2019).

points, then the continuum of the present theory contains too great a manifold of possibilities. I also believe that this too great is responsible for the fact that our present means of description miscarry with the quantum theory" (1916: 379). Nowadays, Joseph Henson (2009), for instance, argues that a lack of short-distance cut-offs prevents us from obtaining the finiteness of the semi-classical black hole entropy (cf. Wüthrich, 2013 for discussion).

We find in the previous paragraph some of the basic intuitions that motivated the development of CST: (i) spacetime continuum is not the ultimate reality, (ii) spacetime continuum emerges from a discrete structure (a collection of discrete spacetime points, called the elements of the causal set), (iii) spacetime points (or events) are related by a partial order, and (iv) this partial order has the meaning of the causality relations between spacetime events. Specifically, CST postulates that the fundamental description of spacetime is not a continuum, but some discrete structure of elementary events ordered by a relation of causality, to which the continuum is only an approximation. To put it another way, CST's main hypothesis is that the spacetime continuum disappears on sufficiently small scales and is superseded by an ordered discrete structure, the causal set (or *causet* for short), the relation of which with the continuum is conceived as one of coarse-grained, macroscopic representation. CST is, in this sense, nothing more than an attempt to show that "[…] at appropriately large scales, this discrete quantum structure approximates[60] the smooth metric manifolds that represent spacetime in general relativity" (Wüthrich, 2013: 227). Furthermore, the *causet* is generally thought as 'growing' as new elements are added one by one to the future of already existing elements. This process of 'growth', which echoes C. D. Broad's temporal becoming, is said to unfold in a 'covariant' manner, such that it seems compatible with relativity. The claim is then that CST (when augmented with a dynamics) allows one to 'rescue' temporal becoming from relativity, and hence to provide a *naturalistic* basis for GBT. In a nutshell, CST stands out from other approaches to quantum gravity (i) by its conceptual and logical simplicity, (ii) by the fact that it incorporates "[…] the assumption of an underlying spacetime discreteness organically and from the very beginning" and (iii) by the fact that it gives rise to a framework "[…] in which time is an active process of 'becoming' that can be identified with the continual birth of new elements of the causal set" (Sorkin, 2006: 1007).

But, what exactly is a *causet*? A *causet* is simply an ordered pair $< C, \leq >$ constituted of a set C of elementary events and a binary relation '$\leq$' on C satisfying the following conditions:

(i) *Reflexivity*: For all $x \in C$, we have $x \leq x$.
(ii) *Anti-symmetry*: For all $x, y \in C$, we have $x \leq y$ and $y \leq x$ implies $x = y$.
(iii) *Transitivity*: For all $x, y, z \in C$, we have $x \leq y$ and $y \leq z$ implies $x \leq z$.
(iv) *Locally finiteness*: For all $x, z \in C$, we have $| \{y \in C \mid x \leq y \leq z\} | < \aleph_0$ (i.e. the cardinality of C has to be less than $\aleph_0$).

[60] As already mentioned (cf. Sect. 4.1), the term 'approximates' is misleading, since the continuum is more fine-grained than the discrete structure.

These simple conditions constitute the basic kinematic assumptions of CST. In particular, the condition of *Anti-symmetry* prevents the *causet* from being the equivalent of closed time-like curves,[61] and the condition of *Locally finiteness* implies that the *causet* is a discrete structure. The relationship induced by the basic order relation '$\leq$' allows for a variety of interpretations. For example, the relationship '$x \leq y$' can variously be described as 'x *precedes* y', 'x is an *ancestor* (or a *parent*) of y', 'y is a *descendant* (or a *child*) of x', 'y lies to the *future* of x', or 'x lies to the *past* of y'. Physically, this ordering can be thought as a microscopic counterpart of the macroscopic '*earlier-later than*' relation. The reason *why* the word 'causal' comes into the picture is that, as has been explained in Sect. 4.2, to say that an event e_1 is earlier than an event e_2 is to say that e_1 could exert a causal influence on e_2. These kinematic assumptions suggest that CST is naturally interpreted in structuralist terms, which corroborates the thesis defended in Sect. 3.4, according to which structures are ontologically primary, while individuals (such as spacetime points, events, and objects) have a mere derivative status. Indeed, the elements of a causet naturally appear as "completely featureless events",[62] while the relation '$\leq$' is "the only concrete physical relation". As Wüthrich puts it: "[…] it is thus evident that causal sets offer what is arguably the most straightforwardly structuralist example of a physical entity postulated by any physical theory" (2013: 233).

What is remarkable is that this structure seems sufficient to reproduce the geometry of four-dimensional spacetime (cf. Reichenbach, 1969; Robb, 1936; Zeeman, 1964).[63] The details would bring us too far, but the main idea is easy to grasp: given that light-cones can be defined in causal terms and that, in the continuum, the light-cones determine the metric up to a conformal rescaling (cf. Sect. 4.2), it appears that (given minimal regularity conditions)[64] the causal order of a Lorentzian manifold captures fully the conformal metric, as well as the topology and the differential structure. To understand this, it is worth remembering that, according to GR, the geometry of spacetime (but not its size) is determined by its causal structure, and hence can be defined by ten numbers to be specified at each spacetime point (cf. Sect. 4.5). These ten numbers correspond to the Einstein tensor, which can be represented schematically as a matrix that possesses four rows and four columns (16 numbers) but that is symmetric in the indices (i.e. along the diagonal), so that it comprises 10 different numbers: four along the diagonal, six on the top right (and six on the bottom left, but they are the same). To capture the geometry of spacetime

[61] As a result, it should be noted that "[…] however causal set theory will relate back to general relativity, it will not be able to reproduce the full theory as general relativity permits time travel in the sense of causal loops" (Wüthrich, 2013: 229, cf. also Smeenk & Wüthrich, 2011).

[62] These elements might properly be considered as *truly* bare particulars (as defined in Sect. 3.10).

[63] This claim is grounded in work by Kronheimer and Penrose (1967), Hawking *and al.* (1976), and Malament (1977), showing that, for a continuum Lorentzian spacetime, "[…] the causal order and local scale information are equivalent to the full geometry" (Dowker, 2020: 147).

[64] Such as the absence of closed causal curves. For more details on this point, see Wüthrich and Callender (2016).

is therefore to capture these ten numbers.[65] Nine of these numbers are given by the light-cones themselves (which trace all the possible light signals that can be sent from, or received, at a certain spacetime point). Given that these light-cones allow for a causal interpretation – for any event *e*, no event that lie outside the absolute past of *e* can be the cause of anything that would influence *e*, and no event outside the absolute future of *e* can be causally influenced by *e* (cf. Sect. 4.2) – it seems that all the information light-cones contain is encoded in the partial ordering of the causet elements. Thus, since a *causet* specifies the causal ordering among events, it defines light-cones and, thereby, provides nine of the ten numbers necessary to describe the geometry of spacetime.

The missing number, which corresponds to spacetime volume, cannot be recovered from the causal order of a Lorentzian manifold. However, as Riemann suggested (see above), in the context of a *discrete* order, the volume can be obtained in another way: "[…] by equating the number of causet elements to the volume of the corresponding region of the spacetime continuum that approximates *C*" (Sorkin, 2002: 7). To get a feel for the numbers involved in such kind of counting, Sorkin (2006) estimates that a region of spacetime of spatial volume 1^3 cm and a temporal extent of 1 sec is composed of around 10^{139} elements. Although it is an incredible number, which might explain *why* the granularity of spacetime has not yet been observed in laboratories, it is still a finite number, which exemplifies the idea that the structure of reality is fundamentally discrete. Thus, the order carries 9/10 information and the number (which corresponds to the volume) 1/10. Together they add up to 10/10 (this is what Sorkin's famous slogan 'order + number = geometry' is meant to express). This underpins the claim that a causal set can indeed be approximated by a Lorentzian geometry: "[…] the causal set's order relation proved the approximating continuums causal order and the local physical scale is set by the causal set's discreteness" (Dowker, 2020: 147).[66]

As has been said, these basic considerations provide the kinematical starting point for a theory of discrete quantum gravity based on causal sets. But, this cannot be the end of the story, since the vast majority of causets sanctioned merely by the kinematic axiom "[…] do not stand in a relation of faithful approximation to spacetimes in low-dimensional manifolds" (Wüthrich & Callender, 2016: 4). This is sometimes called 'the inverse problem'. What is additionally required is a dynamics, which allows one to select, among the great variety of kinematically possible causets, those that are approximated by a relativistic spacetime. The basic idea is that the dynamics should be specified by some further axioms, which are intended to perform the selection: only the causets that satisfy them can successfully be

[65] Considering coordinate freedom redundancy, there are only really six numbers; four can be fixed via coordinate gauge choice.

[66] More precisely, one admits that a spacetime approximates a causal set iff the three following conditions are satisfied: (i) "[…] the causet's causal relations are preserved on the emergent level of the relativistic spacetime", (ii) the local scale is fixed, and (iii) the discrete structure does not "[…] give rise to an emerging spacetime with significant curvature at a scale finer than of the fundamental structure" (cf. Wüthrich & Callender, 2016: 3).

reasonable models of the theory.[67] This is absolutely necessary if one wants to show *how* classical spacetimes emerge and, thereby, *why* GR is as successful as it is. From a phenomenological point of view, this step will turn out to be the most interesting aspect of CST since, taken seriously, the dynamics might help rescue temporal becoming and, therefore, make GBT work in a relativistic spacetime setting. Intuitively, this dynamics depicts a process of 'growth' or 'cosmological accretion', at each step of which an element of the *causet* comes into being. This new element is regarded as the "[…] 'offspring' of a definite set of the existing elements – the elements that form its past" (Sorkin & Rideout 2000: 3). Thus, the process of growth has always been explicitly linked to the passage of time by the proponents of CST. A few lines further on, Sorkin and Rideout say that "[t]he phenomenological passage of time is taken as a manifestation of this continuing growth of the causet. Thus, we do not think of this process as happening 'in time', but rather as 'constituting' time […]" (2000: 3).[68] It is worth noting that, although the process of growth entails a succession of births in a definite order, it does not presuppose any notion of absolute simultaneity (cf. Sorkin, 2007: 156).

Specifically, a popular dynamics by which the growth in the causets might take place is called the 'classical sequential growth dynamics' (CSG) (cf. Rideout & Sorkin, 2000; Varadarajan & Rideout, 2006). CSG is to be understood as a "[…] stochastic process starting from the empty set and adding elements one by one, with the transitions governed by probabilities satisfying a Markov condition[69]" (Earman, 2008: 155). In addition to the Markov condition, CSG also respects the following three conditions: (i) internal temporality (which prevents events from 'birthing' in the past of events that have already become), (ii) discrete general covariance, and (iii) Bell causality. In that sense, CSG assigns probabilities to each transition from every finite causal set to its possible 'children' in accordance with certain physical principles. For example, considering Fig. 4.8, the birth of a can be construed as a transition from the empty causet to the (unique) causet of one element; it occurs with probability 1. The birth of b, however, can occur in two different ways: either b will be a child of a (as depicted in the diagram), or it will not; each of these two

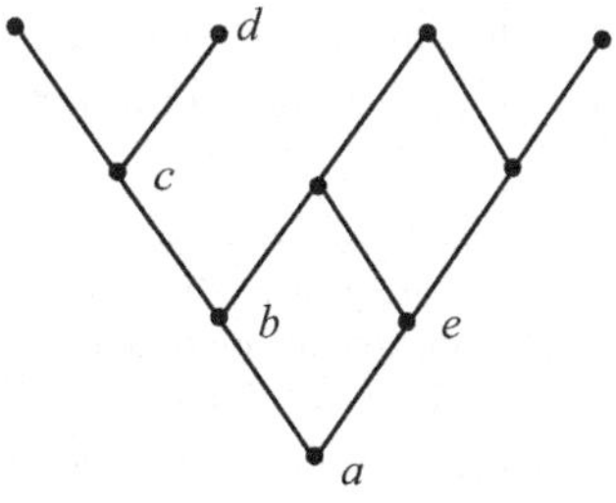

Fig. 4.8 A causet as a Hasse diagram. (A similar figure can be found in Sorkin (2006: §3))

[67] Various axioms can be envisaged here. But, as we will see, proponents of CST generally impose at least the Markov condition, internal temporality, discrete general covariance, and Bell causality.

[68] Accordingly, there is no other meaningful birth-order than the one implied by the relation '$\leq$'.

[69] That means that the probabilities for the transitions "[…] only depend upon the 'initial' and the 'final' state, but not on what transpired before the 'initial' state" (Wüthrich, 2013: 230).

events occur with non-zero probability. At subsequent stages the number of possible causets rises quickly: as Sorkin calculated, after the fourth birth, one can have "[...] any of 16 non-isomorphic causal sets, while after the tenth there are already over two million distinct possibilities (2567284 to be precise)" (2007: 154). The logical space of all these possibilities is delimited by two extreme cases: (i) each new element acquires all the previous ones as ancestors, and the result is a *chain*, a causet that corresponds to one-dimensional Minkowski space, (ii) none of the elements has ancestors (they are space-like separated from each other), and the result is an *anti-chain*, a causet that corresponds to no spacetime whatsoever. In between (i) and (ii), one finds the most interesting cases, where, for example, there are CSG analogs of cyclical cosmologies in such a way that reality grows with successive cycles of collapse and reexpansion.

Although CSG is, as its name indicates, *classical* (which means that no allowance is made for quantum interference between possible distinct transitions from any causal set to its 'children'),[70] a quantum version of the theory is expected within the near future. This would, however, not affect the underlying kinematics or 'ontology' of the theory; as Sorkin makes it clear, "[...] the criterion of 'discrete general covariance' could carry over essentially unchanged from the classical to the quantal case" (2007: 155). In particular, the quantum version of CSG is expected to arise from a decoherence functional (which is itself an "[...] indeterministic, yet mostly irreversible process" (Weinert, 2004: 275)), defined on sets of histories (causal sets).[71] Decoherence functional would thus generalize "[...] the notion of probability measure to allow for interference of distinct possibilities" (Varadarajan & Rideout, 2006: 2). This is good news, since the rejection of determinism has been depicted as a necessary (though non-sufficient) condition for the future to be called 'open' (cf. Sect. 2.5). Of course, conceiving causation in a non-deterministic context is not a straightforward matter[72]; but, as many have insisted,[73] essential quantum mechanical experiments (e.g., the double-slit experiment) suggest that this is required to establish the validity of quantum theory. Furthermore, this suggests that, although the quantum interpretation that at first sight deals better with relativistic causality is actually deterministic, 'going deeper' allows indeterminism to be compatible with relativistic causality, and thus with the open future.

The most exciting aspect of CSG is that it seems to provide an objective correlate of our pre-theoretic intuition that new things are created in the present (cf. *Temporal*

[70] As a consequence, CSG cannot model quantum gravity effects involving interference, which implies that the nature of temporal becoming and the emergence of spacetime in a full quantum version of the approach remain mysterious.

[71] More precisely, a decoherence functional is a complex-valued function of pairs of histories that measures their mutual quantum interference (cf. Dowker & Halliwell, 1992).

[72] According to Laplace, it is even hopeless, since determinism and causation are the same concept. As Weinert puts it: "[...] Laplace bases his superhuman intelligence on the assumption that 'one ought to regard the present state of the world as the effect of its antecedent state and as the cause of the state that is to follow'" (2004: 199).

[73] Cf., for instance, Weinert (2004: 275).

Becoming).[74] And, more generally, the unceasing occurrence of birth-events that build up the causet seems to explain *why* we experience time as we do (e.g., the arrow of time points from past to future, the future is open while the past is fixed, etc.). Furthermore, since CSG is said to unfold in a generally covariant manner, it seems perfectly compatible with relativity. Therefore, not only could CSG restore temporal becoming within physics, but it could do so "[…] without paying the price of a return to the absolute simultaneity of pre-relativistic days" (Sorkin, 2006: §4). In this last respect, CST is preferable to Tooley's theory, which has been shown incompatible with Lorentzian covariance (cf. Sect. 4.4). If CST turns out to be correct, it might therefore have significant consequences for the philosophy of time. In particular, as Sorkin (2007) and Earman (2008) point out, since CST (augmented with CSG) encodes a 'birthing' process akin to C. D. Broad's notion of temporal becoming, this model might underwrite a growing block theory of time (GBT). This appears very plausible for the three reasons that I detail now.

First, CST provides a natural way to construct a 'spatially extended' present, which seems required by both the geometric and the dynamic component of GBT: "[…] the events co-present with the 'here-now' are those events on a space-like slice – technically, a 'maximal antichain' – that is, a maximal set of events such that any two events are incomparable in terms of the relation $\leq$. A sequence of presents would then be a partition of a causet into such maximal antichains" (Wüthrich & Callender, 2016: 5).[75] Second, CST pictures a discrete spatiotemporal substructure which is "[…] four-dimensional from the very beginning, but which at any stage of its growth is still *incomplete*" (Sorkin, 2007: 157). In that sense, at any stage of the process of 'growth' we stop, we are left with the maximal elements of C, which form a 'future boundary' of the growing causet. Third, and more generally, what makes it difficult to express GBT within relativistic settings is that the ontological growth it entails is generally conceived as an accretion of thin layers of Newtonian absolute time, which therefore betrays general covariance. But why should accretion be conceived this way? What the causal set approach to quantum gravity (augmented with CSG) tells us is that the accretion does not proceed with respect to one particular cosmic time function, but rather by the birthing of events through a discrete stochastic process – this highlights that the coming into existence in a particular order is a different issue from a universe wide layer of increase. And, when the outcome of this stochastic process is approximated by a sequence of classical general relativistic spacetimes, the result may (or may not) look like a growing block model of time, as described in the previous chapter.

[74] The question as to whether this form of temporal becoming conforms to C. D. Broad's original intuitions and motivations is discussed in Wüthrich and Callender (2016). This point will be developed below.

[75] However, although this definition of the present looks natural, one will see that it raises some difficulties.

Of course, this approach could be criticized for being too speculative. After all, CST remains in an incomplete stage of development.[76] For example, as has been acknowledged, one of the major aims of current work on CST is to develop an appropriate quantum version of CSG. But, against a widespread opinion, there is at least one important thing that CST teaches us: intuitive phenomena, such as temporal becoming, the passage of time, or the openness of the future, are logically consistent with the four-dimensional Lorentzian manifold of relativity.[77] As, for instance, Fay Dowker puts it: "[…] CSG models are counterexamples to the claim that Relativity implies a block universe view of time" (2020: 148). It is therefore wrong to claim that contemporary physics forces a 'block universe' view on us, since it can accommodate an objective (i.e. mind-independent) form of temporal becoming. Specifically, when a CSG model produces a causet *C*, well-approximated by a continuum spacetime like our universe, a relativistic form of temporal becoming arises, whereas no comparable phenomenon could be found in the 'block universe' view. This suggests that philosophers of physics might have been wrong to neglect our intuitions, which have always spoken in favor of a becoming conception of time. Perhaps some of the elements for understanding the inner structure of time were pre-theoretically given to us, but they have been ignored, because of a misleading conception of science, according to which scientific concepts are completely divorced from our actual experience (cf. Sect. 1.1).

Another criticism concerns the 'spatially extended' present that a maximal antichain within the structure of a causal set may seem to offer. Wüthrich and Callender argue that such a way of constructing the present from the resources of CST is problematic, mainly because, for any given event *here-now*, "[…] there are in general many maximal antichains of which it is an element" (2016: 5). As a consequence, the present cannot be uniquely defined (various sets of co-present events can be created); and privileging one antichain over the others seems just as problematic as privileging one particular foliation in Minkowski spacetime: "[…] a partition of a causal set into such maximal antichains would not be invariant under automorphisms of its structure" (*id*). Supposing that the latter criticism is well-founded, does it undermine the way GBT has been defined in the previous chapter, i.e. as the only *asymmetric* A-theory of time that accepts *Temporal Becoming* (Sect. 3.5)? The answer is 'not necessarily', but surely some adjustments would be needed. For example, the reflection symmetry, which allows one to distinguish between the *symmetric* and the *asymmetric* structures, would have to be operated, not around 'the present' axis (since no unique maximal antichain could play this role), but around a spatiotemporal point. The operation in question is a central (not axial) symmetry;

[76] Nonetheless, as Sorkin (2006) makes clear, CST already has solid achievements, especially regarding its predictions about the so-called cosmological constant. Indeed, supernova observations indicate that gravity (in the current cosmological epoch) ceases to be attractive at large distances; this can be explained by a cosmological constant of precisely the order of magnitude that has been anticipated by CST.

[77] As has been shown in Sect. 4.4, Howard Stein (1968, 1991) had already demonstrated that the possibility of temporal becoming was not ruled out by Minkowski spacetime.

from a fixed point Ω, it transforms any point M into an image point M', such that Ω is the midpoint of the segment [MM']. Further, temporal becoming would have to be conceived as a local (rather than global) phenomenon, sometimes called 'asynchronous becoming' (composed of multiplicity of '*nows*'). Sorkin, for instance, explicitly adopts this conception when he says: "[…] our 'now' is (approximately) local and if we ask whether a distant event space-like to us has or has not happened yet, this question lacks intuitive sense. […] the supposition of [a 'super observer', who would take in all of existence at a glance] would lead to a distinguished 'slicing' of the causet, contradicting the principle that such a slicing lacks objective meaning ('covariance')" (2007: 158).

This obviously leads to a further criticism: the dynamic picture of the world offered by CST (through a CSG model) is exotic and, therefore, does not answer to the typical GBT's demands. Indeed, although a CSG model allows for a (perhaps localized) form of becoming, it rules out the possibility of a single physical world that grows and change, while this seems to be at the heart of C. D. Broad's original project. Oliver Pooley (2013), for instance, claims that Sorkin's view is best understood as a version of non-standard A-theory of time (in Kit Fine's sense), but this kind of view as been shown to be not fully intelligible (cf. Sect. 3.2). However, things might be less dramatic than they seem, especially because Earman (2008), and Wüthrich and Callender (2016) have developed two concurrent *objective* and *global* forms of becoming that seem both respectful of CST-*cum*-dynamics and the structure of relativity, so that the above objection does not seem definitive. But, before introducing these two options, it is worth examining a further challenge that they might face. For that purpose, consider the singleton set, and suppose that it births a time-like separated element, Max's birthday, at label time $l = 1$. Then, suppose that this two-element causet births a third element, Mary's birthday, which is space-like separated from the other two elements, at label time $l = 2$. This is path α. By contrast, path β births Mary's birthday space-like separated from the singleton set, and *then* births Max's birthday, which is time-like separated only from the singleton set – the situation is depicted in Fig. 4.9. General covariance (which is used as a condition to derive the dynamics) requires that the probability of any particular causet arising is *independent* of the path to get to that causet and, therefore, that "[…] the product of the transition probabilities along the links of α is the same as that for β (and any other such path)" (Wüthrich & Callender, 2016: 13).

This situation is sufficient to make apparent a difficulty: according to relativity, there is no fact of the matter as to which event – either Max or Mary's birthday – came first. As Wüthrich and Callender put it: "[t]o say which one happened 'first' is to invoke non-relativistic concepts" (2016: 13). To put it another way, the world grows from C_1 (the singleton set) to C_2 to C_3, but there is not a determinate fact as to whether C_2 consists of the singleton plus Max's birthday or the singleton plus Mary's birthday. General covariance entails that, if a causal set can be reached via two different paths (and hence can occur with two inequivalent labellings), the transition probabilities along each path must be the same. In that sense, it is often said that general covariance ensures that the labels used in the growth process are 'pure

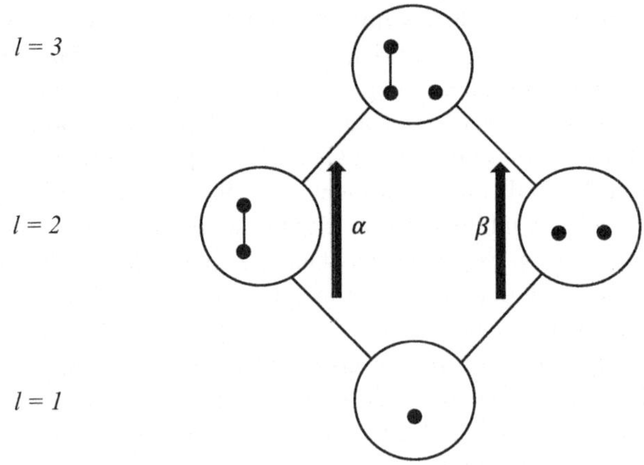

Fig. 4.9 Stochastic sequential growth. (A similar figure can be found in Earman (2008: 157), and Wüthrich & Callender (2016: 14))

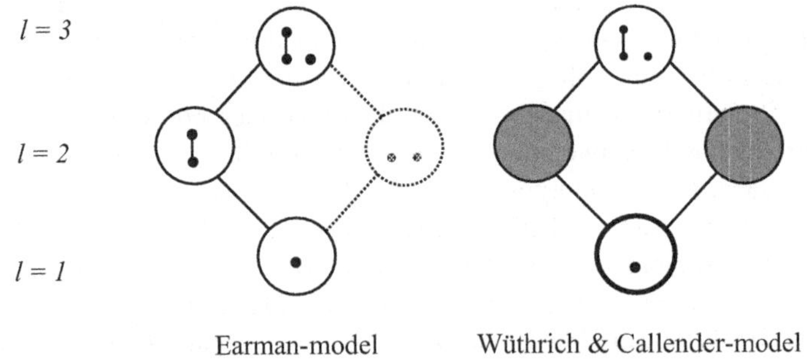

Fig. 4.10 Earman's distinction between actualized and non-actualized alternatives, and Wüthrich and Callender's bizarre form of indeterminacy

gauge'. This simple example shows that conceiving a physically consistent growth (within the causets) happening in time is not a straightforward matter. Fortunately, there seem to be promising solutions on the market. For instance, John Earman (2008) assumes (against a many-worlds interpretation of the stochastic birthing process) that only one of the possible paths, α or β, is actualized. Accordingly, in Fig. 4.9, the becoming of the actual world is modeled by path α (say), but not by path β, since the causet on the right is not actualized. As Earman puts it: "[s]ince quantum aspects are ignored in classical sequential growth dynamics, it seems fair [...] to assume that each stage only one of the possible alternatives is actualized" (2008: 158). This 'philosophical' addition to the causets allows one to regain an objective and global form of becoming (cf. Fig. 4.10, on the left).

Of course, one might be reluctant to add such a hidden variable moving up the causet, especially because this move seem to betray the 'standard' interpretation of the labels as 'pure gauge'.[78] If so, another option seems available: "[t]here simply is no determinate fact as to whether C_2 contains *a* or *b*; but there is a determinate fact that it contains one of them" (Wüthrich & Callender, 2016: 15). In other words, whereas it is determinate that C_2 consists either of the singleton *plus* Max's birthday or of the singleton *plus* Mary's birthday, it is indeterminate whether C_2 has Max's birthday in it, and it is indeterminate whether C_2 has Mary's birthday in it. This kind of indeterminacy should remind us *how* Elizabeth Barnes and Ross Cameron (2009, 2011) characterize the openness of the future (and borderline cases): there are multiple determinate (precise) states between which the world is unsettled (cf. Sect. 2.7). The logic is the same here: there are two possibilities, C_2-*cum*-Max's birthday, and C_2-*cum*-Mary's birthday, and the world is unsettled as to which one obtains. In that sense, indeterminacy should not be regarded as a mere epistemic phenomenon, but as a metaphysical phenomenon, which may concern the extension of a (finite) causet (i.e. *what* elements are to be found in it), but not its cardinality (i.e. *how many* elements are to be found in it). If this idea is coherent, "[…] CST does permit a new kind of – admittedly, radical and bizarre – temporal becoming" (Wüthrich & Callender, 2016: 15): things get determinate at every finite stage of becoming. For instance, in Fig. 4.9 (on the right), all the 'ancestors' of C_3 must have determinately been obtained, with the notable exception of its two immediate ancestors. At this stage, C_3 is neither determinate nor indeterminate, since it has not yet come into existence (the two structures below are therefore not at the same stage of development). Then, at the next stages (not shown in Fig. 4.9), C_3 comes into existence, and one of its two immediate ancestors – it is indeterminate which one – gets determinate. And so on.

In a nutshell, GBT implies an increase in ontology: new thin layers come successively into being. Admittedly, if these layers correspond to slices of Newtonian absolute time, GBT is at odds with contemporary physics. But *why* should one think of the layers in such an old-fashioned way? CST allows for a birthing of events through a discrete stochastic process, which has to be quantal to generate a truly manifold-like causet. Although we do not possess such a quantal dynamics yet, we can imagine *how* it will formally look like: Varadarajan and Rideout (2006), for instance, claim that it will be expressed in terms of a decoherence functional. The outcome of this process, when it is approximated by a sequence of general relativistic spacetimes, may resemble a hypersurface Becoming model. Classical relativistic spacetime would, in that sense, be an emergent feature of temporal becoming (though the quantal nature of this process remains mysterious). Although temporal becoming within CST is often regarded as a local phenomenon (cf. Sorkin, 2007), and therefore poses a threat to the intuitive picture of a single physical world that grows and changes, Earman (2008) and Wüthrich and Callender (2016) show that CST allows for global forms of temporal becoming. In particular, they found some

[78] As Earman admits: "[i]f it is legitimate to worry that the actualization/non-actualization distinction presupposes an external time, then the ability of the causets approach to deliver a model of Becoming is seriously undermined. But I see no reason for the causets proponents to give in to this worry […]" (2008: 158).

generally covariant senses in which one could say that there is becoming, but not to be conceived as a 'asynchronous' or frame-relative phenomenon, but as an absolute phenomenon. Although this phenomenon could simply be expressed by the fact that cardinality of the causets grows, Earman (2008), Wüthrich and Callender (2016) went looking for more interesting facts about causets: either by distinguishing between actualized and non-actualized alternatives (Earman), or by putting forward a form of indeterminacy about what elements are to be found in the causets (Wüthrich & Callender). Taking these propositions seriously, it seems that CST might offer a naturalistic basis to GBT (at least as defined in the previous chapter). And, since GBT is well-designed to accommodate some of our basic intuitions about time (e.g., the future is open, while the past is fixed), CST might play a crucial role in the reconciliation of the manifest image and contemporary science.

4.7 Reconciling GBT with Science Fiction: The Case of Time-Travel

In this section, the proposal is to move from science to science fiction by considering the possibility of time-travel. Although this possibility became a real possibility with the advent of SR and GR, the expression 'science fiction' seems justified, since no such travel has yet been undertaken by any of our contemporaries. The possibility of time-travel has always captured the popular imagination; one can find hundreds of books and movies which explore it. The best example is perhaps H. G. Wells' *The Time Machine*, in which the protagonist leaves his own time (1895) to travel in to the distant future and then return to his own present. Of course, some philosophers refuse to take these scenarios seriously, mainly because travels into the future involve certain oddities (e.g., a mathematician can bring back from the future a mathematical proof that he decides to publish in the present), and travels into the past involve paradoxes (e.g., the grand-father paradox). But, as has been argued in the second chapter (cf. Sect. 2.4), these difficulties may find some solution through an appropriate treatment of the question of 'What we can and cannot do in the future or the past?' (cf. Lewis, 1976). The possibility of time-travel has already been considered in the present book, especially in Sect. 2.8, where it was argued that dynamic branching-tree models cannot allow for it. Arguably, presentism is in no better position, since it recognizes no non-present location to which to travel (but see Dowe, 2000, and Baron & Miller, 2019: §8). This argument, which highlights the incompatibility of presentism with the possibility of time-travel, is commonly called the 'no destination' argument. It might therefore seem that if one can show that GBT allows for the possibility of time-travel, it would have a certain advantage over at least both of these two competing theories. It is worth noting that, in what follows, the question of the compatibility of GBT and time-travel will be asked on a non-relativistic conception of time / spacetime, and with no reference to the 'bare particular' view developed in the third chapter (since it is unclear whether it allows for time-travel). Furthermore, this section will offer the opportunity to address the thorny question of persistence: 'How can things exist at different moments of time within GBT?'. It will be shown that GBT is *a priori* compatible with both endurantism, i.e. the view that things are wholly present whenever they exist, and perdurantism (or four-dimensionalism), i.e. the view that things have temporal parts (or stages).

At first sight, GBT appears to be better positioned than presentism when it comes to accommodating the possibility of time-travel. Specifically, GBT partially resists to the 'no destination' argument: although GBT *a priori* excludes time-travel to the future (since it implies that future temporal locations are unreal), it offers various past locations to which to travel. In particular, given GBT, if t_5 is the objective present, then there seems no reason to suppose that one could not travel to *all* temporal locations that are in the past relative to t_5. Yet, Kristie Miller contests the latter claim; she argues that GBT (just as presentism) is incompatible with backward time-travel, because "[…] it requires that present states be caused by non-existent indeterminate future states" (2005: 229). Let me explain. As usually conceived, time-travel is ruled by (at least) the following three principles: (i) *P* is a genuine time-traveler only if all of *P*'s temporal parts are united by some causal relation, (ii) it is not possible to change the past (cf. Sect. 2.4), and (iii) it is not possible to travel from a non-existent location to an existing one (cf. Miller, 2005: 227). The first principle, which demands a causal continuity among the stages of a time-traveler, is intended to rule out cases of counterfeit time-travel: if Fred is randomly created by a demon at a time *t* and it happens to be a duplicate of a stage of Sam destroyed at a later time t_5, this should not count as a case of time-travel (cf. Lewis, 1976: 148). The second principle implies that "[…] if one can travel to some past location t, it will be true at all times subsequent to t, that one existed at t. And it will be true at t, that one exists at t" (Miller, 2005: 227). In that sense, if Max does not exist at *t* when *t* is the present, then Max did not exist at *t* when *t* is in the past (and, therefore, Max has not traveled to *t*). The third principle is the converse of the 'no destination' argument. It is partly justified by the fact that a future non-existent event allegedly cannot be the cause of a current event (e.g., a time-traveler existing *now*), particularly given that it is supposed to be indeterminate whether the future event in question will occur (cf. Miller, 2005: 228).

Now, suppose that t_5 is the objective present, and at t_5 Max travels back in time to t_1. Since it is not possible to change the past, "[i]f [Max] exists at t_1 when t_1 is the objective past, then [Max] must exist at t_1, when t_1 is the objective present. [But, according to GBT] when t_1 is the objective present, t_5 does not exist. So [given that it is not possible to travel from a non-existent location to an existing one], it is not possible for any time traveller to have travelled from t_5 to t_1" (Miller, 2005: 229). To put it another way, it seems that, given GBT, there are only two ways things could turn out: either (i) Max does not exist at t_1 when t_1 is the objective present, and hence he does not exist at t_1 when t_1 is the objective past (since it is not possible to change the past), or (ii) Max does exist at t_1 when t_1 is the objective present, but he is not a time-traveler, since his t_1 temporal part cannot be causally connected to any temporal part that exists in the future (such a temporal part does exist, after all). Neither of these options allows for time-travel. It might therefore be concluded that GBT is incompatible with this possibility.

This objection may seem powerful, but only if one assumes the Lewisian conception of time-travel, according to which time-travelers are *perduring entities* that have temporal parts at their point of departure (as well as at any other time at which they exist). One can therefore distinguish two potential options to resist the objection: (i) to reject perdurantism in favor of endurantism (time-travelers do not have temporal parts; they are wholly present whenever they exist), and (ii) to accept a revised conception of perdurantism which allows one to say that, in some sense, Max exists at t_1 when t_1 is present, and has a future temporal part that *will* exist at t_2. The first option is supported by orthodoxy: perdurantism does not go well with presentism, and for similar reasons, it might seem that perdurantism does not go well with GBT either. After all, why should one claim that an object has temporal parts at other times than the present (perdurantism) if these parts do not exist? Of course, a presentist could say that these temporal parts *existed* and exist no longer, but in what sense would they be parts of the object? It might seem, as Trenton Merricks puts it, that "[a]n object cannot have another object as a part if that other object does not exist" (1995: 524).[79] To put it another way, according to perdurantism (at least in its classical conception), an object is an aggregate of *all* its temporal parts while, given non-eternalist ontologies, there might be times at which such an aggregate does not exists. For example, since the Eiffel Tower will probably still be standing tomorrow, growing blockers have every reason to believe that some of its temporal parts do not exist (but will exist). It might therefore seem preferable for growing blockers (and other non-eternalists) to reject temporal parts and, thus, to endorse endurantism.[80]

However, although this first option allows one to reject the second principle of time-travel scenarios, it is not clear that it is a way out for growing blockers, since this principle can be replaced by another one, upon which an 'endurantist' version of Miller's argument can be built: "[…] P is a genuine time traveller only if for every times t and t* at which P exists, there is some causal relation that holds between P at t and P at t*" (Miller, 2005: 225). A better option might therefore be to argue that the classical conception of perdurantism (or four-dimensionalism) is misleading. Of course, if to perdure an object must *exist at different times* by having parts at those times, then perdurance is at odds with non-eternalist ontologies (at least regarding objects that will be partially located in the future). But, it seems coherent to say that

[79] As we will see, Merrick's principle can be challenged. For example, Sally Haslanger gives the following counterexample: "[...] my maternal grandmother is part of my extended family even though she does not (presently) exist" (2003: 325).

[80] This first option may not sound very plausible since, assuming that GBT is true, it faces with the problem of multi-location (an object enjoys multi-location just in case it is wholly present at more than one (distinct) spacetime region). Specifically, endurantism implies that each object must be wholly located within each time it exists, so that if past and future times exist (as growing blockers assume), some objects are multi-located, which might lead to paradoxes (cf. Barker & Dowe, 2003). However, many philosophers (e.g., Eagle, 2016; McDaniel, 2003) have argued that there is nothing logically or conceptually problematic about multiple-location.

a persisting object consists of its present and past parts, and of those parts that it *will* have in the future (*pace* Merricks). As, for instance, Lawrence Lombard (1999) points out, one must carefully distinguish between two senses of 'exist' if one is a perdurantist: (i) the 'straightforward sense' in which instantaneous temporal parts exist at a time, and (ii) the 'derivative sense' in which an object, i.e. a whole composed of all of its temporal parts, exists at some time. According to (i), if temporal parts exist at a certain time, they exist at this time *entirely* (they are three-dimensional entities) and they have all of their (spatial) parts at this time. According to (ii), objects (e.g., material objects, people, etc.) exist at some time in virtue of having a temporal part that does; but one is enough, it does not need to have all of their parts at this time. It is obviously this second, derivative, sense of 'exist' that is the interesting one for perdurantists; the first one being accepted by everyone: "[…] if there are any three-dimensional instantaneous entities, it is uncontroversial that they exist entirely at the time they do" (Benovsky, 2007: 84).

Criticizing the classical conception of perdurantism, Lombard argues that "[…] what is obvious is only that an object that exists *at a time*, *t*, cannot have, *at t*, another object as a part, if that other part does not exist *at t*. But what the perdurantist wishes to say is *not* inconsistent with that. What, in [the straightforward] sense, exists now – e.g., the present temporal part of a computer – is something that does not (ever) have as parts anything that does not exist now. But what exists now in [the derivative] sense – the computer – is something that does (at some time or other) have parts that do not exist now; but what exists now in that [derivative] sense does not now have those parts" (1999: 256). In short, an object, such as a computer, construed as composed of temporal parts (perdurantism), exists now in the derivative sense of having a part that exists now (in its entirety). Of course, it might be objected that, since what exists now, in the derivative sense, does not have its non-present temporal parts now, it does not, if eternalism is false, have them at all (cf. Merricks's principle). This objection rests on the idea that objects must have their temporal parts in the same way that they have their spatial parts: temporal parts, like spatial parts, must exist in their entirety. But this does not apply to temporal parts in general. As Berit Brogaard observes: "[…] events are commonly understood as having temporally extended parts even though these never exist as a whole but only though their successive stages" (2000: 346). Similarly, although an object exists at *t* in virtue of having some temporal parts (perdurantism), it is not required that all of its temporal parts exist at *t*. The existence of an object at *t* (in the derivative sense) merely requires that one of its temporal parts exists at *t* (in the straightforward sense). Fabrice Correia and Sven Rosenkranz support a similar idea when they write: "[m]ereological fusions, if they exist, exist whenever, and wherever, one of their parts exist. […] residents of spacetime may, for all that, be mereological fusions of spatiotemporal parts, as long as one of their spatiotemporal part is located here-now" (2018: 151).

In the light of the distinction between the straightforward and derivative senses of 'exist', Merricks's principle appears clearly problematic. On one reading, in which 'exist now' means 'exists in its entirety at the present time' (the straightforward sense), the principle is inconsistent with the existence of entities that have temporal parts. On a second reading, in which 'exist now' means 'now has a temporal part that exists in its entirety at the present time' (the derivative sense), the principle is simply false (cf. Lombard, 1999: 257). Once again, what exists now, in the straightforward sense, is a temporal part that does not have as parts anything that does not exist now; what exists now, in the derivative sense, is an object (material object, people, etc.) that does have parts (temporal parts) that do not all exist now, but this object does now have those parts. This revised conception of perdurantism allows one to conclude that the argument for the incompatibility of GBT and time-travel fails. To return to the previous example, Max can travel from t_5 to t_1 since, although his t_5 temporal part does not exist when t_1 is the objective present (in the straightforward sense), he still has it as a part, i.e. he still exists partially in virtue of having his t_5 temporal part in the future. Max can therefore be a genuine time-traveler, even if GBT is true, since there is a sense in which Max is composed of all of his parts (including the future ones), one of them (his t_5 temporal part) being the reason why he is now visiting t_1 when t_1 is the objective present. Does this undermine (i) the general causal picture and (ii) the openness of the future, as Miller (2005) claims? The answer seems to be 'no'. What matters is that (i) there will be a cause at t_5 such that it will explain *why*, at t_1, there was a time-traveler coming from the future (which seems guaranteed by the fact that Max partially exists in virtue of having a t_5 temporal part in the future), and (ii) this cause is not *predetermined* (nothing there is or was, in conjunction with how it is or was, makes Max's time-travel inevitable): it was not inevitable for Max to visit the past.

By the way, taking McTaggart's conception of change seriously (the only way in which events can *genuinely* change is by first being future, then present and finally past), it seems that this revised conception of perdurantism is capable of avoiding what is generally introduced as the main objection against perdurantism (at least in its four-dimensionalist version), namely that it entails a changeless world. The objection runs as follows. Consider an apple that is green at t_1 and brown at t_2. What this amounts to, according to four-dimensionalists, is that one of the apple's temporal parts is green, and another is brown. However, when considering this account of change, it might be objected that what we are looking for is an account of how a *single object* (the apple) can change, while four-dimensionalists are telling us a story about *different objects* (different temporal parts) having different properties. In the four-dimensionalist picture, "[w]hat we have is not change of an individual, but replacement of one changeless object (one temporal part) by another changeless one" (Benovsky, 2007: 81). Specifically, instead of saying that the apple has changed from being green to being brown between t_1 and t_2, four-dimensionalists say that the t_1 temporal part of the apple has changelessly the property of being green and the t_2 temporal part of the apple has changelessly the property of being brown. Since the apple itself cannot lose or gain any such properties, it seems that four-dimensionalism leaves no room for genuine change. As Peter Simons puts it: "[...] Lewis's favoured four-dimensional alternative is not an explanation of change but an elimination of it, since nothing survives the change which has the contrary properties" (2000: 65).

Now, it seems that the revised conception of perdurantism might allow us to overcome the elimination of change by bringing back the passage of time into the four-dimensional picture. In particular, combined with GBT (rather than eternalism), perdurantism becomes the view according to which (i) objects have temporal parts, and (ii) at any given time, only past and present temporal parts of objects exist (in the straightforward sense). Of course, just as four-dimensionalism, GBT-perdurantism entails that "[…] an object does not gain or lose properties; rather, different properties are possessed by different [temporal parts]" (Brogaard, 2000: 348). But, contrary to classical-perdurantism, GBT-perdurantism entails that new temporal parts are coming into existence – and this in a way that seems to capture our most basic intuitions according to which change has taken place. Specifically, a change of *x* has taken place if and only if "[…] (i) there is an entity *z* which is a present [temporal part] of *x*; and (ii) there was an entity *y* which was a previous [temporal part] of *x*; and (iii) *z* has a different set of [intrinsic] properties than *y* had" (Brogaard *id*). The creation of new temporal parts *plus* new intrinsic properties (e.g., being red, being square, etc.) seems thus sufficient for a change to take place. Moreover, GBT-perdurantism definitely seems better suited to account for the asymmetry between the 'open future' and the 'fixed past' than the classical alternative. Classical-perdurantism implies that it is settled that people have the temporal parts that they have; these parts (including future ones) exist *tenselessly*, after all. By contrast, on GBT-perdurantism, whereas it is settled that people have the temporal parts that they in fact had and now have, it might be unsettled what parts people will have in the future (at least assuming that physical determinism is false), because these parts do not exist yet.

4.8 Conclusion

Physics informs and frames the metaphysical debate on the nature of time. In (neo-) Newtonian mechanics, time (as well as space) is absolute; it is regarded as an empty container of discrete, atomistic *nows*, which cannot be affected by any material agency. This reflects the intuitive idea that the world evolves *in* time in an objective manner. The (neo-)Newtonian picture allows for a four-dimensionalist interpretation, according to which spacetime can be foliated into three-dimensional hyperplanes. (Neo-)Newtonian mechanics therefore admits an absolute notion of objective simultaneity: it allows one to establish a common time system for a group of people (even if these people are moving with respect to each other), so that everyone in this group will agree on which events are simultaneous. This makes the (neo-)Newtonian mechanics a friendly environment for expressing GBT, provided that the layers of existence successively coming into being are seen as slices of Newtonian absolute time.

However, the (neo-)Newtonian mechanics has been shown to be deficient both *theoretically* (cf. Maxwell's equations of electromagnetism) and *experimentally* (cf. the Michelson-Morley outcome). It has therefore been replaced with relativistic physics, which recognizes that the notion of absolute simultaneity is unfounded. This makes GBT notoriously at odds with relativity since, as the Putnam-Rietdijk argument points out, the unreality of the future requires an objective notion of absolute simultaneity. Fortunately, many options – both compatibilist and incompatibilist – are available to growing blockers to escape the pressure exerted by the Putnam-Rietdijk argument. Compatibilist options consist of either challenging some of the premises of the argument (Sklar, 1974), or re-conceptualizing GBT to make it expressible in relativistic settings (Correia & Rosenkranz, 2018; Stein, 1968). Incompatibilist options consist of either rejecting the metaphysical relevance of SR (Bourne, 2006; Zimmerman, 2008), or denying that SR is approximately true of the world (Tooley, 1997). Although both families of options come with consequences, those of the compatibilist options seem more problematic. Specifically, the compatibilist options seem in tension with both our intuitions (e.g., the intuition that there is single universe-wide border between the past and the future), and some considerations from quantum mechanics (e.g., experimental results connected with John Bell's theorem). This is what led us to privilege an incompatibilist option in the remainder of the chapter.

The incompatibilist option in question took the form of a quantum theory of gravity, the causal set theory, which, since it encodes a 'birthing' process akin to the notion of temporal becoming, might give a second wind to GBT. This 'birthing' process (understood as a discrete stochastic process) is expressed by a dynamics, the so-called 'classical sequential growth dynamics' (CGS), which is in charge of selecting, among the great variety of kinematically possible causal sets, those that are approximated by a relativistic spacetime. Although a version of CGS that incorporates quantum aspects still has to be developed, this approach promises to offer a naturalistic basis to GBT, which was so far regarded as purely speculative. Although, becoming within CST is often regarded as a local phenomenon (cf. Sorkin, 2007), Earman (2008), and Wüthrich and Callender (2016) have shown that CST permits global forms of temporal becoming, which allows one to preserve the intuitive picture of a single physical world that grows and changes. Taking their propositions seriously, CST may play a predominant role in the reconciliation of the manifest image and contemporary science (at least supposing that it is to be successful as a theory of quantum gravity). Finally, it has been shown that GBT is (at least in principle) compatible with time-travel scenarios, provided that one accepts, for instance, a revised conception of perdurantism.

Chapter 5
Conclusion

Abstract In this conclusion, I first briefly recall the theoretical framework within which the present book was undertaken: an attempt at reconciliation of the manifest image with contemporary science. Second, I recapitulate the main obtained results; I especially insist on the fact that an ontological characterization of the intuitive asymmetry between the 'open future' and the 'fixed past' (Chap. 2), as it can be accommodated by a specific version of GBT (Chap. 3), might find some support in nascent approaches to quantum gravity (Chap. 4). Finally, I consider some new directions that could be pursued. In particular, I consider some practical significance that GBT may have with respect to moral and emotional concerns.

5.1 An Attempt at Reconciliation

In the philosophy of time, one generally opposes two traditions of thought: there are the subjectivists who are mostly concerned with how time is commonly experienced (e.g., phenomenologists, philosophers of inner life), and the scientists who are mostly interested in time as science, especially physics, describes it (e.g., empiricists, naturalist metaphysicians). However, as it was argued (Sect. 1.1), this opposition rests on a caricatural picture of science, according to which scientific concepts are completely divorced from our actual experience. A paradigmatic example of this idealization of science is Ladyman and Ross (2007, 2013), who famously claim that "[a]ttaching epistemic significance to metaphysical intuitions is anti-naturalist [...]" (2007: 10). A more realistic view recognizes that human experience and intuitions play an important role in science, at least in the sense of background assumptions influencing how scientific data are interpreted (cf. Morganti & Tahko, 2017). Taking this latter view seriously, it seems that a third way of doing philosophy of time can be envisaged: reconciling the time of human experience with that of contemporary science. This mainly requires developing a framework within which pre-theoretic and scientific data can be articulated non-paradoxically. The present book was written in accordance with this general philosophical scheme.

V. Grandjean, *The Asymmetric Nature of Time*, Synthese Library 468,
https://doi.org/10.1007/978-3-031-09763-8_5

The main purpose of this book was to account for the intuitive asymmetry between the 'open future' and the 'fixed past'. This asymmetry was mainly, but not exclusively, illustrated by examples drawn from our practical life: there are things we can do to affect how the future will unfold (e.g., finding a cure for cancer, acting in an ethically responsible manner), whereas the past is beyond our control (e.g., Napoleon definitely lost the battle of Waterloo). Against some form of skepticism brought by science (especially by contemporary physics), it was argued that this aspect of human experience reflects *how* time truly is: unlike space, time is of an asymmetric nature. To make that case, the book was divided into three main parts; each of them devoted to a separate methodological step: (i) the characterization of the asymmetry, (ii) the accommodation of the asymmetry through a model of the temporal structure of the world, and (iii) the reconciliation of this model (and hence the asymmetry) with our best physics. From a broader perspective, (i), (ii), and (iii) brought a framework within which human experience and science both contribute to a better understanding of our familiar, but paradoxically elusive, concept of time.

Specifically, in the second chapter, I reviewed various philosophical characterizations of the asymmetry between the 'open future' and the 'fixed past'. I especially explored the confrontation between two types of characterizations: *perspectival* (e.g., semantic, epistemic, anthropocentric) and *substantial* (e.g., physical, modal, metaphysical, ontological). In the third chapter, I proposed a new way of introducing the A-theoretic models of the temporal structure of the world, based on some relevant geometric features, and I figured out which of these models is best designed to accommodate the asymmetry in openness between the future and the past. In this chapter, I answered questions such as 'How the existence in the past should be conceived?' and 'Why the existence in the past differs sharply from the existence in the present?'. In the fourth chapter, I wondered whether fundamental physics, as many philosophers would have us believe, undermines any attempt to defend an open-future view. To address this question, I first introduced both (neo-)Newtonian and Einsteinian basics. Then, while (neo-)Newtonian mechanics appears to be a friendlier environment, I wondered whether A-theoretic models could be expressed in relativistic settings. Finally, I looked for some process of 'growth' (in which new events come into existence) in nascent approaches to quantum gravity, especially in the causal set theory (CST), in order to restore the manifest image. In the last section, I moved from science to science fiction to wonder whether GBT is in principle compatible with time-travel scenarios.

5.2 The Summary of the Results

In the second chapter, I argued that the intuitive asymmetry between the 'open future' and the 'fixed past' reflects a *fundamental* (rather than *derivative*) phenomenon that can only be observed in a non-deterministic world, where the future history is *not* nomologically necessitated by the current history. More specifically, I argued that this phenomenon should be characterized as a kind of *worldly unsettledness* that is to be expressed in terms of *underdetermination* (rather than *overdetermination*): there being facts of the matter about what happened, but not about what will happen. This characterization was contrasted with some form of skepticism, mainly supported by scientific arguments, which aim to show that no asymmetry is to be found within the 'fundamental features' of reality. After all, the 'block universe' view of time, favored by physicists, is *isotropic* (spacetime has no intrinsic direction), and the fundamental laws of physics are *time-reversal invariant* (they do not distinguish the future-direction from the past-direction). These scientific arguments lead to think that the asymmetry is merely perspectival (i.e. it arises from the peculiar way our minds interact with reality), and should therefore be characterized in semantic, epistemic or even anthropocentric terms.

However, as it was shown, perspectival characterizations of the asymmetry raise at least two issues, although they may be relevant to explain our having certain practices. First, they seem illegitimate: the fact that science fails to capture certain phenomena does not entail that these phenomena are merely perspectival. For example, there are dozens of natural phenomena (e.g., northern lights, will-o'-the-wisps) that, although science had for long regarded as subject-dependent, turned out to be objective features of reality. Second, assuming that the open future merely reflects some sort of human ignorance about what will happen, it has to be explained *where* this ignorance comes from, while no convincing explanation has been found yet. Most attempts involve the second law of thermodynamics. For example, Jenann Ismael claims that: "[w]hat explains our greater knowledge of the past than the future is that along that [thermodynamic] gradient [produced by the rise of entropy], inferences from the present, surveyable macroscopic state of the world to its past [...] are much more powerful inferences than inferences from the present to the future" (2016: 143). But this kind of Reichenbach-inspired explanation at best postpones the problem: given that there is no directedness in fundamental physics, where does the thermodynamic asymmetry in time come from?

In the third chapter, I argued that the problem of the asymmetry in openness between the future and the past should partly be conceived as the problem of establishing a difference in some of the geometric properties of our spatiotemporal models. Accordingly, I proposed a reformulation of the A-theories of time, which are traditionally introduced as various answers to the ontological question 'Do the future and the past exist?', in geometric terms. In particular, I distinguished two kinds of A-theories of time, the *symmetric* and the *asymmetric* theories, which differ with respect to whether, when reflection symmetry is operated around 'the present' axis, the outcome is an unchanged or a transformed spatiotemporal structure. More specifically, a symmetric theory is a theory such that possibly always the structures it describes is reflection invariant; conversely, an asymmetric theory is a theory such that necessarily sometimes the structure it describes is not reflection invariant. In that respect, eternalism and presentism were called 'symmetric', whereas GBT and SBT were called 'asymmetric' A-theories of time. Then, in order to distinguish between the various forms symmetric and asymmetric A-theories may adopt, a new question was introduced: 'Is temporal becoming (i.e. the creation of new things in the present) real?'. One consequence of this proposal is that GBT is no longer to be seen as an ill-conceived hybrid between two polar opposites (eternalism and presentism), but as a real alternative: it is the only *asymmetric* A-theory that accepts *Temporal Becoming*. This new characterization revealed GBT to be better positioned than its rivals to accommodate various past-future time asymmetries, including the asymmetry between the 'open future' and the 'fixed past'. In particular, unlike presentism, GBT can avail itself of the 'no fact of the matter' account of the openness of the future, while keeping the past fixed. Moreover, assuming that *physical determinism* is false, GBT implies, through *Temporal Becoming*, that new things are created in the present, while (at least) some of them are not made inevitable by how things located in the present or the past of now are or were. GBT thus meets the two necessary and sufficient conditions for regarding the future and the past as open and fixed, respectively.

However, despite these attractive features, GBT is often criticized for not being a viable alternative to presentism and eternalism, because of (i) the epistemic objection, according to which GBT would lead to absolute skepticism about *where* we are temporally located, and (ii) the apparent contradiction between GBT and relativistic physics, especially with respect to absolute simultaneity. In the second half of the third chapter, I addressed the epistemic objection, by showing that it relies on a mistaken assumption, namely that the reality of the past entails that events are occurring in the past. Specifically, I argued that the past should be conceived as a spatiotemporal region where nothing occurs, since it is exclusively populated by 'bare particulars' that are responsible for the continuity of existence of both continuants (people, tables, planets, etc.) and occurrents (events, processes, etc.), through the passage of time. In that sense, becoming past involves alteration in things' intrinsic properties to such an extent that they cease to belong to their *natural kind* (rejection of natural kind essentialism). But, this alteration does not cause the things to cease to exist (rejection of *Annihilation*), since they now correspond to bare particulars, which are at least freed from all the properties that made these things belong to the natural kind to which they belonged when present (e.g., the property of *occurring*, if the things in question were events). Finally, assuming that if one's belief is occurring, then one introspectively knows it, I concluded that we (as constituents of conscious events) can be confident of being right when we think that the time at which we exist is the objective present; no such conscious event (or any other event) could occur in the past, after all.

In the fourth chapter, I treated the objection based on the apparent incompatibility between GBT and relativistic physics. I first explained what this incompatibility is about, by contrasting Newtonian mechanics with the Special theory of relativity (SR). Then, I showed that, since SR does not privilege any way of slicing four-dimensional Minkowski spacetime into three-dimensional hyperplanes, it challenges the two components – geometric and dynamic – at work in the definition of GBT. Specifically, SR does not allow one to pick out a hyperplane as being (i) the unique 'present' axis around which reflection symmetry can be operated, and (ii) a universe-wide border where new events come into existence. Faced with this objection, I classified the possible reactions into two families of options: the compatibilist and the incompatibilist options. Whereas the compatibilist options take the metaphysical relevance of SR for granted, the incompatibilist options contest it. I then argued that the incompatibilist options are more promising, since compatibilist options conflict with both our common intuitions (e.g., the intuition that there is single universe-wide border between the past and the future) and some considerations from quantum mechanics (e.g., experimental results connected with John Bell's theorem).

In the second half of the fourth chapter, I argued that the General theory of relativity is of no help to the compatibilist strategy, since (i) the possibility of 'closed time-like curves' renders GBT even less plausible, and (ii) relativistic causality puts GR in no better position than SR when accommodating quantum considerations (e.g., the violations of John Bell's inequality, the collapse of the wave-packet). Taking that for granted, I suggested that a naturalistic basis for GBT (if any) should rather to be found in the nascent theories of quantum gravity. In that respect, I considered the causal set theory (CST), which rests on the idea that continuum spacetime disappears on sufficiently small scales, and is superseded by an ordered discrete structure: the causal set. This structure seems sufficient to reproduce the geometry of four-dimensional spacetime. Although CST is still in its infancy, it promises to underpin a growing block model of time, which was so far regarded as purely speculative. For, CST, when augmented with the 'classical sequential dynamics' (CSG), which takes the form of a discrete stochastic process of 'growth', provides an objective correlate of the intuitive notion of temporal becoming. Since this dynamics is said to unfold in a generally covariant manner, it renews the hope of reconciliation between GBT and relativity. Whereas temporal becoming within CST is often regarded as a local phenomenon, Earman (2008), and Wüthrich and Callender (2016) showed that CST permits global forms of temporal becoming that allow one to preserve the intuitive picture of a single physical world that grows and changes. CST might therefore be the missing link between the manifest image and science. Assuming indeterminism, it preserves the possibility of being both a scientific realist *and* a defender of the view that the future is open. Finally, I showed that GBT is in principle compatible with time-travel scenarios, provided that one accepts, for instance, a revised conception of perdurantism.

5.3 Future Directions

The question of the nature of time has intrigued philosophers for centuries, but much work still remains to be done, especially with regard to the reconciliation between the manifest and the scientific images. Whereas the present book was mainly concerned with the intuitive asymmetry between the 'open future' and the 'fixed past', many other aspects of human experience of time need to be reconciled with contemporary physics. For example, one can think of time passage, the spatially-extended present, and enduring objects, which all *a priori* seem to conflict with what relativistic physics tells us of the world. It would therefore be interesting to apply the three-step methodology – characterization, modeling, reconciliation – to these aspects too, in order to get a more comprehensive picture of the temporal structure of the world. Second, it would be desirable to give further consideration to the conception of the past – 'the bare past' – developed in the third chapter. One could, for instance, wonder whether this conception is compatible with time-travel, or whether bare particulars are sufficient to underpin the complexity of the past (what about past properties, or past instants?). Third, whereas the second law of thermodynamics has been shown *insufficient* to explain intuitive temporal asymmetries, one still needs an account that articulates GBT with the increase in entropy. Perhaps a starting point could be to insist, as Maudlin (2007) does, that a *one-way* temporal evolution is required to express the increase of entropy away from a low-entropy. Then, an idea would be to show that this temporal evolution is packed into *Temporal Becoming*, which is constitutive of GBT.

Another direction would be to explore the practical significance of GBT. It indeed seems that GBT may help clarify our thinking about matters of moral and emotional concern. For example, consider our fear of dying. Death is traditionally defined as the "[…] unequivocal and permanent end of our existence" (Nagel, 1979: 61). As L. W. Sumner puts it: "[t]he death of a person is the end of that person; before death he *is* and after death he *is not*. To die is therefore to cease to exist" (1976: 153). This definition highlights what frightens us the most about death: our own annihilation (cf. Luper-Foy 1987).[1] Of course, death will always be a matter of fear, but GBT could persuade us that things are less bad than they seem. After all, GBT removes from death its existential significance: even if we become something intrinsically different after death (perhaps a bare particular!), we definitely remain something. In that sense, death does not bring absolute annihilation: lives do not cease to exist, they simply have a beginning and an end with, hopefully, creation of durable good in between. Accordingly, the loss of being is not something that should be feared, since the permanentness of being is guaranteed by GBT. The most we have to lose, so to speak, is our experience of life (or something closely related), which undoubtedly sounds less tragic. In short, one could say that the past is existence without afterlife. By contrast, the birth of a child (which is literally to be understood as the coming into the world of a new human being) should always be a matter of great rejoicing.

Closely related, GBT might provide a partial explanation as to *why* we care more about future experiences than past ones (especially when the experiences in question are of the painful and pleasure variety). This emotional asymmetry[2] can be illustrated by the fact that, in general, we would prefer that bad things (e.g., a headache, an awkward meeting) be in the past rather than in the future (cf. Hare, 2013: 507). Of course, there have been valiant attempts to account for this asymmetry. Arguably the most influential one is due to David Lewis (1979b), who argues that wanting bad things to be in the past involves wanting of myself (construed as a

[1] It is worth noting that Epicurus makes use of this definition to show that it is irrational to fear death. As he puts it: "[…] as long as we exist, death is not with us; but when death comes, then we do not exist" (1940: 31).

[2] This asymmetry is known as 'the temporal value asymmetry' (cf. Caruso et al., 2008).

person-stage) that I be to-the-future of bad things. But, this account leads to arbitrary considerations. As Caspar Hare puts it: "[w]hy should I want my pains to be in the past if wanting pain to be in the past just amounts to wanting myself (construed as a person-stage) that I be to the future of pain? Wouldn't that be just like standing in a row of soldiers [...], knowing that one of them has a toothache, and wanting of myself that I be to the south of the pain?" (2013: 514). It indeed seems that Lewis' account fails to explain *why* a desire that I (construed as a person-stage) be to the future-of-pain is less arbitrary than a desire that I be to the south of pain.

Interestingly, GBT might offer a partial solution to this problem. Let me sketch the general idea. The most plausible reason *why* we care more about future experiences than past ones is that we are typically both intrigued and scared by the unknown. As H. P. Lovecraft puts it at the very beginning of his classical essay on fear and the supernatural: "[t]he oldest and strongest emotion of mankind is fear, and the oldest and strongest kind of fear is the fear of the unknown" (1974: 13). In the same vein, Elias Canetti writes that "[t]here is nothing that man fears more than the touch of the unknown" (1962: 15). My wanting my pain to be past could therefore involves my preferring to have a detailed rather than a poor knowledge about what happens to me. Taking this idea seriously, it seems that GBT, at least when enriched with the 'no fact of the matter' account of openness (cf. Sect. 2.9), is well positioned to account for this emotional asymmetry. Indeed, since this theory provides an immediate explanation as to *why* future experiences remain largely unknown (we cannot know more about the future than what is settled about it), it also helps explain *why* future experiences are a matter of great concern (in comparison to past experiences).

At this point, it could be objected that it is simply wrong to assume that we *always* attach more importance to future events than to past ones. For example, it might seem that a person condemned to death could, just before his execution, care less about his future than about his past (since his future is limited to a few minutes). Arguably, this person would have no more plans or hopes, whereas he would remember the pivotal moments of his life, and perhaps feel some regrets about his past crimes. Yet, it seems that GBT cannot account for this case, and hence neglects the variety of our emotional concerns. In response, two claims can be made. First, in the above case, it is not clear that the person cares more about the past than the future. In such a situation, one can rather be inclined to think that he would be *obsessed* by his approaching death and other future events. He might ask questions such as 'Will I suffer?', 'What is there after death?', 'Will my children have a happy life?' or 'Will my wife get remarried?'. Second, even assuming that the person sentenced to death mostly cares about past events, this can perfectly be accommodated by GBT. As a reminder, GBT (alone) does not entail that the future is open, nor does it entail that the future is a matter of great ignorance. For example, in a fully deterministic context, what will happen is fixed (no matter whether the future exists or not) and, therefore, everything is predictable (cf. Laplace [1814] 1951). So, if a person sentenced to death does not attach great importance to the future, this could easily be explained by the fact that *he believes* that *his* future is fully determinate (there is no way for him to avoid his execution – no escape plan, no legal recourse, etc.). He perfectly *knows* what is going to happen, and therefore his future contains no more mystery he might fear. But, here again, these are just naïve thoughts; a perspicuous account would require further investigations, especially within epistemology and the philosophy of emotions.

Of course, these future directions are merely intended as suggestions: one may choose to change them to match our own philosophical purposes; the future is open, after all.

Bibliography

Adams, R. M. (1974). Theories of actuality. *Noûs., 8*, 211–231.

Adams, R. M. (1979). Primitive thisness and primitive identity. *Journal of Philosophy., 76*, 5–26.

Adams, R. M. (1981). Actualism and thisness. *Synthese, 49*, 3–41.

Adams, R. M. (1986). Time and thisness. *Midwest Studies in Philosophy., 11*, 315–329.

Albert, D. (1994). *Quantum mechanics and experience*. Harvard University Press.

Albert, D. (2000). *Time and chance*. Harvard University Press.

Albert, D. (2015). *After physics*. Harvard University Press.

Alcoloumbre, T. (2001). Création ou Eternité. *Pardès, 31*, 73–82.

Andreoletti, G. (2020). Back to the (Branching) Future. *Acta Analytica, 35*, 181–194.

Anscombe, G., & Elizabeth, M. (1956). Aristotle and the Sea-battle. *Mind, 65*, 1–15.

Aristotle. (1995). *Politics*. E. Barker (Trans.). Oxford University Press.

Aristotle. (1999). *Physics*. R. Waterfield (Trans.). Oxford University Press.

Aristotle. (2014). *Categories, on interpretation, and on sophistical refutations*. E. M. Edghill (Trans.), Neeland Media LLC.

Armstrong, D. (1978a). Nominalism and realism. In *Nominalism and realism: Universals and scientific realism*. Cambridge University Press.

Armstrong, D. (1978b). A theory of universals. In *Nominalism and realism: Universals and scientific realism*. Cambridge University Press.

Armstrong, D. (1989). *A combinatorial theory of possibility*. University Press.

Armstrong, D. (1997). *A world of states of affairs*. Cambridge University Press.

Armstrong, D. (2004). How do particulars stand to universals? In D. Zimmerman & K. Bennett (Eds.), *Oxford studies in metaphysics* (Vol. 1, pp. 139–154).

Arsenijevic, M. (2016). Truth and the open future: the solution to Aristotle's sea battle challenge with the principle of bivalence retained. *Unspecified.*

Atkins, P. W. (1986). Time and dispersal: The second law. In R. Flood & M. Lockwood (Eds.), *The nature of time* (pp. 80–98). Blackwell.

Augustine. (2006). *The confessions*. M. P. Foley (ed.). Hackett Publishing Company.

Aydede, M. (2019). Pain. *Stanford Encyclopedia of Philosophy*. E. Zalta (ed.). https://plato.stanford.edu/entries/pain/

Bäck, A. (1992). Sailing through the Sea-Battle. *Ancient Philosophy., 12*, 133–151.

Bacon, A. (2019). Is reality fundamentally qualitative? *Philosophical Studies., 176*, 259–295.

Bailey, A. (2012). No bare particulars. *Philosophical Studies., 158*, 31–41.

Baker, L. R. (2010). Temporal reality. In J. Campbell, M. O'Rourke, & H. Silverstein (Eds.), *Time and identity* (pp. 27–47). MIT Press.

Bardon, A. (Ed.). (2012). *The future of the philosophy of time*. Routledge.

V. Grandjean, *The Asymmetric Nature of Time*, Synthese Library 468,
https://doi.org/10.1007/978-3-031-09763-8

Barker, S., & Dowe, P. (2003). Paradoxes of multi-location. *Analysis, 63*, 106–114.

Barnes, E. (2010). Ontic vagueness: A guide for the perplexed. *Noûs, 44*, 601–627.

Barnes, E. (2013). Metaphysically indeterminate existence. *Philosophical Studies., 166*, 495–510.

Barnes, E. (2014). Fundamental indeterminacy. *Analytic Philosophy, 55*, 339–362.

Barnes, E., & Cameron, R. (2009). The open future: Bivalence, determinism and ontology. *Philosophical Studies, 146*, 291–309.

Barnes, E., & Cameron, R. (2011). Back to the open future. *Philosophical Perspectives, 25*, 1–26.

Barnes, E., & Cameron, R. (2017). Are there indeterminate states of affairs? No. In E. Barnes (Ed.), *Current controversies in metaphysics*. Routledge.

Baron, S. (2012). Presentism and causation revisited. *Philosophical Papers., 41*, 1–21.

Baron, S. (2017). Back to the unchanging past. *Pacific Philosophical Quarterly., 98*, 129–147.

Baron, S., & Miller, K. (2019). *An introduction to the philosophy of time*. Polity Press.

Barovan, S., & Colyvan, M. (2016). Time enough for explanation. *Journal of Philosophy., 113*, 61–88.

Bartels, A., & Wohlfarth, D. (2014). How fundamental physics represents causality. In M. Galavotti, D. Dieks, W. Gonzalez, S. Hartmann, T. Uebel, & M. Weber (Eds.), *New directions in the philosophy of science* (pp. 485–500). Springer.

Baxter, D. (2001). Instantiation as partial identity. *Australasian Journal of Philosophy., 79*, 449–464.

Beall, J. C. (2012). Future contradictions. *Australasian Journal of Philosophy., 90*, 547–557.

Bell, J. L. (2013). Continuity and infinitesimals. In E. Zalta (Ed.), *Stanford Encyclopedia of Philosophy*. https://plato.stanford.edu/entries/continuity/

Belnap, N. (1991). Before refraining: Concepts for agency. *Erkenntnis, 34*, 137–169.

Belnap, N. (1992). Branching space-time. *Synthese, 92*, 385–434.

Belnap, N. (2001). Double time references: Speech-act reports as modalities in an indeterminist setting. In F. Wolter, H. Wansing, M. de Rijke, & M. Zakharyaschev (Eds.), *Advances in modal logic* (pp. 37–58). World Scientific Publishing Co. Pte. Ltd.

Belnap, N. (2003). Agents in branching space-times. *Journal of Sun Yatsen University, Social Science Edition, 43*, 147–166.

Belnap, N. (2005). Branching histories approach to indeterminism and free will. In B. Brown & F. Lepage (Eds.), *Truth and probability essays in Honour of Hugues Leblanc* (pp. 197–211).

Belnap, N., & Green, M. (1994). Indeterminism and the thin red line. *Philosophical Perspectives, 8*, 365–388.

Belnap, N., Perloff, M., & Xu, M. (2001). *Facing the future: Agents and choices in our indeterministic world*. Oxford University Press.

Belot, G. (2013). Time in classical and relativistic physics. In H. Dyke & A. Bardon (Eds.), *A companion to the philosophy of time* (pp. 185–200). Wiley-Blackwell.

Bennett, J. (1984). Counterfactuals and temporal direction. *Philosophical Review, 93*, 57–91.

Bennett, K. (2011). By our Bootstraps. *Philosophical Perspectives, 25*, 27–41.

Benovsky, J. (2007). On presentist perdurantism. *Sats: Nordic Journal of Philosophy., 8*, 79–88.

Benovsky, J. (2008). The bundle theory and the substratum theory: Deadly enemies or twin brothers? *Philosophical Studies, 141*, 175–190.

Benovsky, J. (2011). Endurance and time travel. *Kriterion, 24*, 65–72.

Benovsky, J. (2013). Branching and (In)determinism. *Philosophical Papers, 42*, 151–173.

Bergmann, G. (1967). *Realism: A critique of Brentano and Meinong*. University of Wisconsin Press.

Bergson, H. ([1907] 1944). *Creative Evolution*. A. Mitchell (Trans.). Dover Publications.

Bergson, H. (1972). *Mélanges*. A. Robinet (ed.). PUF.

Bergson, H. (2010). *Time and free will, an essay on the immediate data of consciousness*. F. L. Pogson (Trans.), Charleston: Nabu Press.

Besson, C., & Hattiangadi, A. (2013). The open future, bivalence and assertion. *Philosophical Studies, 167*, 251–271.

Bigaj, T. (2015). Dissecting weak discernibility of Quanta. *Studies in History and Philosophy of Modern Physics, 50*, 43–53.

Bigelow, J. (1996). Presentism and properties. In J. Tomberlin (Ed.), *Philosophical perspectives X, metaphysics* (pp. 35–52). Blackwell.
Bird, A. (2004). Strong necessitarianism: The nomological identity of possible worlds. *Ratio, 17*, 256–276.
Bird, A. (2007). *Nature's metaphysics: Laws and properties*. Oxford University Press.
Bird, A., & Tobin, E. (2017). Natural Kinds. In E. Zalta (Ed.), *Stanford Encyclopedia of Philosophy*. https://plato.stanford.edu/entries/natural-kinds/
Black, M. (1952). The identity of indiscernibles. *Mind, 61*, 153–164.
Blake, R. M. (1925). On Mr. broad's theory of time. *Mind, 34*, 418–435.
Boltzmann, L. (1887). *Vorlesungen über die Principe der Meckanik*. Barth.
Boltzmann, L. ([1886] 1974). The Second Law of thermodynamics. In B. McGuinness (Ed.), *Theoretical physics and philosophical problems* (pp. 13–32). D. Riedel.
Boltzmann, L. ([1898] 2003). *Lectures in gas theory*. Dover.
Borghini, A., & Torrengo, G. (2013). The metaphysics of the thin red line. In F. Correia & A. Iacona (Eds.), *Around the tree* (pp. 105–125). Springer.
Bourne, C. (2002). When Am I? A tense time for some tense theorists? *Australasian Journal of Philosophy., 80*, 359–371.
Bourne, C. (2006). *A future for presentism*. Oxford University Press.
Braddon-Mitchell, D. (2004). How do we know that it is now now? *Analysis, 64*, 199–203.
Braddon-Mitchell, D., & Nola, R. (Eds.). (2008). *Conceptual analysis and philosophical naturalism*. MIT Press.
Braüner, T., Hasle, P., & Øhrstrøm, P. (1998). Ockhamistic logics and true futures of counterfactual moments. In *Proceedings of Fifth International Workshop on Temporal Representation and Reasoning* (pp. 132–139). Sanibel Island, IEEE Press.
Braüner, T., Hasle, P., & Øhrstrøm, P. (2000). Determinism and the origins of temporal logic. In H. Barringer & al. (Eds.), *Advances in temporal logic* (pp. 185–206). Kluwer Academic Publishers.
Briggs, R., & Forbes, G. A. (2012). The real truth about the unreal future. In *Oxford studies in metaphysics* (Vol. 7, pp. 257–304). Oxford University Press.
Briggs, R., & Forbes, G. A. (2017). The growing-block: Just one thing after another? *Philosophical Studies, 174*, 927–943.
Briggs, R., & Forbes, G. A. (2019). The future, and what might have been. *Philosophical studies, 176*, 505–532.
Broad, C. D. (1923). *Scientific thought*. Kegan Paul, Trench, Trubner & Co., Ltd.
Broad, C. D. (1939). Arguments for the existence of God. *Journal of Theological Studies, 40*, 157–167.
Broad, C. D. (1959). A reply to my critics. In P. A. Schilpp (Ed.), *The philosophy of C. D. Broad*. Tudor Publishing Company.
Broadie, S. (1987). Necessity and deliberation: An argument from the *De Interpretatione 9. Canadian Journal of Philosophy., 17*, 289–306.
Brogaard, B. (2000). Presentist four-dimensionalism. *The Monist, 83*, 341–354.
Brown, R. (1991). Divine omniscience, immutability, Aseity and human free will. *Religious Studies, 27*, 285–295.
Brown, H. (2005). *Physical relativity*. Oxford University Press.
Brown, H., & Uffink, J. (2001). The Origins of time-asymmetry in thermodynamics: The Minus First Law. *Studies in the History and Philosophy of Modern Physics, 32*, 525–538.
Burgess, J. (1979). Logic and time. *Journal of Symbolic Logic, 44*, 566–582.
Burgess, J. (1980). Decidability for branching time. *Studia Logica, 39*, 203–218.
Burke, M. (1996). Tibbles the Cat: A Modern "Sophisma". *Philosophical Studies, 84*, 63–74.
Busch, J. (2003). What structures could not be. *International Studies in the Philosophy of Science, 17*, 211–225.
Butterfield, J. (1984). Seeing the present. *Mind, 93*, 161–176.

Butterfield, J. (2007). Reconsidering relativistic causality. *International Studies in the Philosophy of Science, 21*, 295–328.
Butterfield, J. (2013). On time in quantum physics. In H. Dyke & A. Bardon (Eds.), *A companion to the philosophy of time* (pp. 220–241). Wiley-Blackwell.
Byrne, C. (2018). *Aristotle's science of matter and motion*. University of Toronto Press.
Callender, C. (2000). Shedding light on time. *Philosophy of Science, 97*, 587–599.
Callender, C. (2001). Thermodynamic asymmetry in time. *Stanford Encyclopedia of philosophy*. E. Zalta (ed.). https://plato.stanford.edu/entries/time-thermo/
Callender, C. (2008). The common now. *Philosophical Issues., 18*, 339–361.
Callender, C. (Ed.). (2011). *The Oxford handbook of philosophy of time*. Oxford University Press.
Calosi, C., & Wilson, J. (2019). Quantum metaphysical indeterminacy. *Philosophical Studies, 176*, 2599–2627.
Cameron, R. (2014). Parts generate the whole, but they are not identical to it. In A. J. Cotnoir & D. L. M. Baxter (Eds.), *Composition as identity* (pp. 90–108).
Cameron, R. (2015). *The moving spotlight: An essay on Time & Ontology*. Oxford University Press.
Cameron, R. (2016). On characterizing the Eternalism/presentism and Actualism/Possibilism debates. *Analytic Philosophy, 57*, 110–140.
Campbell, K. (1990). *Abstract Particulars*. Blackwell.
Canetti, E. ([1960] 1962). *Crowds and power*. The Viking Press.
Cao, T. Y. (2003). Structural realism and the interpretation of quantum field theory. *Synthese, 136*, 136: 3–24.
Carnap, R. (1931). Überwindung der Metaphysik durch Logische Analyse der Sprache. *Erkenntnis, 2*, 219–241.
Carnap, Rudolf, (1963) The philosophy of *Rudolf Carnap*. P. A. Schilpp (ed.), : Open Court.
Cartwright, N. (2008). *The dappled world: A study of the boundaries of science*. Cambridge University Press.
Caruso, E., Gilbert, D., & Wilson, T. (2008). Asymmetric valuation of past and future events. *Psychological Science, 19*, 796–801.
Casati, R., & Torrengo, G. (2011). The not so incredible shrinking future. *Analysis, 71*, 240–244.
Caulton, A., & Butterfield, J. (2012). On kinds of indiscernibility in logic and metaphysics. *British Journal for the Philosophy of Science, 63*, 27–84.
Chalmers, D. (2014). Intuitions in philosophy: A minimal Defense. *Philosophical Studies, 171*, 535–544.
Chisholm, R. (1981). Time and temporal demonstratives. In K. Weinke (Ed.), *Logik, Ethik und Sprache* (pp. 31–36). R. Oldenburg Verlag.
Chisholm, R. (1990a). Events without times: An essay on ontology. *Noûs, 24*, 413–428.
Chisholm, R. (1990b). Referring to things that no longer exist. In J. Tomberlin (Ed.), *Philosophical perspectives: Vol. 4 (Action theory and philosophy of mind)* (pp. 545–556).
Christensen, F. (1974). McTaggart's paradox and the nature of time. *The Philosophical Quarterly, 24*, 289–298.
Clifton, R., & Hogarth, M. (1995). The definability of objective becoming in Minkowski spacetime. *Synthese, 203*, 355–387.
Cohen, S. M. (1996). *Aristotle on nature and incomplete substance*. Cambridge University Press.
Coles, P., & Ellis, G. (1997). *Is the universe open or closed? The density of matter in the universe*. Cambridge University Press.
Connolly, N. (2011). How the dead live. *Philosophia, 39*, 83–103.
Connolly, N. (2015). Yes: Bare Particulars! *Philosophical Studies, 172*, 1355–1370.
Copeland, J. (Ed.). (1996). *Logic and reality. Essays on the legacy of Arthur prior*. Clarendon Press.
Corish, D. (2009). Could time be change? *Philosophy, 84*, 219–232.
Correia, F., & Iacona, A. (Eds.). (2013). *Around the tree: Semantic and metaphysical issues concerning branching and the open future*. Springer.
Correia, F., & Rosenkranz, S. (2011). *As time goes by: Eternal facts in an ageing universe*. Mentis.

Correia, F., & Rosenkranz, S. (2012). Eternal facts in an ageing universe. *Australasian Journal of Philosophy, 90*, 307–320.

Correia, F., & Rosenkranz, S. (2013). Living on the brink, or welcome Back, growing block! *Oxford Studies in Metaphysics, 8*, 333–335.

Correia, F., & Rosenkranz, S. (2015). Presentism without Presentness. *Thought, 4*, 19–27.

Correia, F., & Rosenkranz, S. (2018). *Nothing to come: A defense of the growing block theory of time*. Springer.

Correia, F., & Rosenkranz, S. (2020a). Temporal existence and temporal location. *Philosophical Studies, 177*, 1999–2011.

Correia, F., & Rosenkranz, S. (2020b). Unfreezing the spotlight: Tense realism and temporal passage. *Analysis, 80*, 21–30.

Correia, F., & Schnieder, B. (Eds.). (2012). *Metaphysical grounding: Understanding the structure of reality*. Cambridge University Press.

Craig, W. L. (1988). *The problem of divine foreknowledge and future contingents from Aristotle to Suarez*. E.J. Brill.

Craig, W. L. (1991). *Divine foreknowledge and human freedom: The coherence of theism*. E. J. Brill.

Craig, W. L. (2000). *The tensed theory of time*. Kluwer.

Craig, W. L. (2001a). *Time and the metaphysics of relativity*. Kluwer Academic Publishers.

Craig, W. L. (2001b). McTaggart's paradox and the temporal solipsism. *Australasian Journal of Philosophy, 79*, 32–44.

Crisp, T. (2003). Presentism. In M. Loux & D. Zimmerman (Eds.), *The Oxford handbook of metaphysics* (pp. 211–245). Oxford University Press.

Crisp, T. (2004). On presentism and triviality. In D. Zimmerman (Ed.), *Oxford studies in metaphysics* (Vol. 1, pp. 15–20). Oxford University Press.

Crivelli, P. (2004). *Aristotle on truth*. Cambridge University Press.

Curtis, B., & Robson, J. (2016). *A critical introduction to the metaphysics of time*. Bloomsbury.

Cushing, J., Fine, A., & Goldstein, S. (1996). *Bohmian mechanics and quantum theory: An appraisal*. Kluwer.

Dainton, B. (2010). *Time and space* (2nd ed.). Routledge.

Darby, G. (2010). Quantum mechanics and metaphysical indeterminacy. *Australasian Journal of Philosophy, 88*, 227–245.

Darby, G., & Pickup, M. (2020). Modelling deep indeterminacy. *Synthese, 198*, 1685–1710.

Davidson, D. ([1970] 2011). Mental Events. *Essays on actions and events* (pp. 207–224). Oxford University Press.

Deasy, D. (2017). What is presentism. *Noûs, 51*, 378–397.

Deasy, D. (2019). The triviality argument against presentism. *Synthese, 196*, 3369–3388.

Denby, D. (2014). Essence and intrinsicality. In R. Francescotti (Ed.), *Companion to intrinsic properties* (pp. 87–109). De Gruyter.

Deng, N. (2010). 'Beyond A- and B-time' reconsidered. *Philosophia, 38*, 741–753.

Deng, N. (2017). Making sense of the growing block view. *Philosophia, 45*, 1113–1127.

Derosset, L. (2013). Grounding explanation. *Philosophers' Imprint, 13*, 1–26.

Diekemper, J. (2004). Temporal necessity and logical fatalism. *Proceedings of the Aristotelian Society, 104*, 289–296.

Diekemper, J. (2005). Presentism and ontological symmetry. *Australasian Journal of Philosophy, 83*, 223–240.

Diekemper, J. (2007). B-theory, fixity and fatalism. *Noûs, 41*, 429–452.

Diekemper, J. (2009). Thisness and events. *Journal of Philosophy., 5*, 255–276.

Diekemper, J. (2013). The existence of the past. *Synthese, 191*, 1085–1104.

Dieks, D., & Redei, M. (Eds.). (2006). *The ontology of Spacetime*. Elsevier.

Dolev, Y. (2006). How to square a non-localized present with special relativity. In D. Dieks & M. Redei (Eds.), *The ontology of Spacetime* (pp. 177–190). Elsevier.

Dolev, Y., & Roubach, M. (Eds.). (2016). *Cosmological and psychological time*. Springer.

Dorato, M. (1995). *Time and reality: Spacetime physics and the objectivity of temporal becoming*. CLUEB.

Dorato, M. (2006a). The irrelevance of the Presentist/Eternalist debate for the ontology of Minkowski Spacetime. In D. Dieks & M. Redei (Eds.), *The ontology of Spacetime* (pp. 93–110). Elsevier.

Dorato, M. (2006b). Absolute becoming, relational becoming and the arrow of time. *Studies in History and Philosophy of Modern Physics, 37*, 559–576.

Dorato, M. (2008). Putnam on time and special relativity: A long journey from ontology to ethics. *European Journal of Analytic Philosophy, 4*, 51–70.

Dowe, P. (2000). The case for time travel. *Philosophy, 75*, 441–451.

Dowker, F. (2020). Being and becoming on the road to quantum gravity, or, the birth of a baby is not a baby. In N. Huggett, K. Matsubara, & C. Wüthrich (Eds.), *Beyond Spacetime: The foundations of quantum gravity* (pp. 143–153). Cambridge University Press.

Dowker, F., & Halliwell, J. (1992). Quantum mechanics of history: The Decoherence functional in quantum mechanics. *Physical Review D, 46*, 1580–1609.

Dowker, F., Henson, J., & Sorkin, R. (2004). Quantum gravity phenomenology, Lorentz invariance and discreteness. *Modern Physics Letters A, 19*, 1829–1840.

Downing, P. B. (1959). Subjunctive conditionals, time order, and causation. *Proceedings of the Aristotelian Society, 59*, 125–140.

Dummet, M. (2012). The place of philosophy in European culture. *European Journal of Analytic Philosophy, 8*, 14–23.

Dummett, M. (1975). Wang's Paradox. In R. Keefe & P. Smith (Eds.), *Vagueness: A reader* (pp. 99–118). MIT Press.

Dummett, M. (1986). Causal loops. In R. Flood & M. Lockwood (Eds.), *The nature of time* (pp. 135–169). Blackwell.

Dyke, H., & Bardon, A. (Eds.). *A companion to the philosophy of time*. Wiley-Blackwell.

Eagle, A. (2016). Multiple location defended. *Philosophical Studies, 173*, 2215–2231.

Earman, J. (1967). Irreversibility and temporal asymmetry. *Journal of Philosophy, 64*, 543–549.

Earman, J. (1974). An attempt to add a little direction to 'the problem of the direction of time. *Philosophy of Science, 41*, 15–47.

Earman, J. (1986). *A primer on determinism*. D. Riedel.

Earman, J. (1992). Determinism in the physical sciences. In J. Earman & C. N. Glymour (Eds.), *Introduction to the philosophy of science* (pp. 232–268). Hackett Publishing.

Earman, J. (2002). What time reversal invariance is and why it matters. *International Studies in the Philosophy of Science, 16*, 245–264.

Earman, J. (2006). The past hypothesis: Not even false. *Studies in History and Philosophy of Modern Physics, 37*, 399–430.

Earman, J. (2008). Reassessing the prospects for a growing block model of the universe. *International Studies in the Philosophy of Science, 22*, 135–164.

Eckhardt, W. (2006). Causal time asymmetry. *Studies in History and Philosophy of Modern Physics, 37*, 439–466.

Eddington, A. ([1928] 1932). *The nature of the physical world*. Cambridge University Press.

Eddington, A. (1935). *New pathways in science*. Cambridge University Press.

Eddington, A. (1939). *The philosophy of physical science*. Cambridge University Press.

Egan, A. (2007). Epistemic modals, relativism and assertion. *Philosophical Studies, 133*, 1–22.

Einstein, A. (1905). Zur Elektrodynamik bewegter Körper. *Ann. Phys., 322*, 891–921.

Einstein, A. (1911). Über den Einfluss der Schwerkraft auf die Ausbreitung des Lichtes. *Ann. Phys., 35*, 99–108.

Einstein, A. (1916). Letter to Walter Dällenbach, Nov. 1916', item 9-072 translated and cited in 'Einstein and the quantum: Fifty years of struggle. In R. G. Colodny (Ed.), *From quarks to quasars, philosophical problems of modern physics* (pp. 379–381). Pittsburgh University Press.

Einstein, A. (1920). *Relativity: The special and general theory*. Routledge.

Ellis, G. (2006). Physics in the real universe: Time and Spacetime. *General Relativity and Gravitation, 38*, 1797–1824.
Ellis, G. (2013). The arrow of time and the nature of Spacetime. *Studies in History and Philosophy of Modern Physics, 44*, 242–262.
Epicurus. (1940). Letter to Menoeceus. In W. J. Oates (Ed.), *The stoic and epicurean philosophers*. Random House.
Esfeld, M. (2004). Quantum entanglement and a metaphysics of relations. *Studies in History and Philosophy of Modern Physics, 35*, 601–617.
Esfeld, M. (2019). Against the disappearance of Spacetime in quantum gravity. *Synthese*. Special Issue on 'Spacetime functionalism', 1–15.
Esfeld, M., & Lam, V. (2008). Structural realism about space-time. *Synthese, 160*, 27–46.
Eva, B. (2018). Emerging (In)Determinacy. *Thought, 7*, 31–39.
Evans, G. (1985). Does tense logic rest on a mistake? In G. Evans (Ed.), *Collected papers* (pp. 343–363). Oxford University Press.
Faye, J. (1989). *The reality of the future: An essay on time, causation and backward causation*. Odense University Press.
Feynman, R. (1982). Simulating physics with computers. *International Journal of Theoretical Physics, 21*, 467–488.
Fine, K. (1975a). Critical notice of Lewis, counterfactuals. *Mind, 84*, 451–458.
Fine, K. (1975b). Vagueness, truth and logic. *Synthese, 54*, 235–259.
Fine, K. (1994). Essence and modality. *Philosophical Perspectives, 8*, 1–16.
Fine, K. (1995). Ontological Dependence. *Proceedings of the Aristotelian Society, 95*, 269–290.
Fine, K. (2001). The question of realism. *Philosopher's Imprint, 1*, 1–30.
Fine, K. (2005). Tense and reality. *Modality and tense: Philosophical papers* (pp. 261–320). Oxford University Press.
Fischer, J. M. (1994). *The metaphysics of free will*. Blackwell.
Fischer, J. M. (2011). Foreknowledge, freedom, and the fixity of the past. *Philosophia, 39*, 461–474.
Fischer, J. M., & Todd, P. (Eds.). (2015). *Freedom, fatalism, and foreknowledge*. Oxford University Press.
Fischer, J. M., Kane, R., Pereboom, D., & Vargas, M. (2007). *Four views on free will*. Blackwell.
Forrest, P. (2004). The real but dead past: A reply to Braddon-Mitchell. *Analysis, 64*, 358–362.
Forrest, P. (2005). General facts, physical necessity and the metaphysics of time. In D. Zimmerman (Ed.), *Oxford studies in metaphysics* (Vol. 2). Clarendon Press.
Forrest, P. (2006). Uniform grounding of truth and the growing block theory. *Analysis, 66*, 161–163.
Forrest, P. (2010). The identity of Indiscernibles. In E. Zalta (Ed.), *Stanford Encyclopedia of philosophy*. https://plato.stanford.edu/entries/identity-indiscernible/
Fraser, J. T., Haber, F. C., & Müller, G. H. (Eds.). (1972). *The study of time*. Springer.
Frege, G. (1982). On sense and reference. In P. Geach & M. Black (Eds.), *Translations from the philosophical writings of Gottlob Frege*. Blackwell.
Frege, G. (1984). Thoughts. In B. McGuinness (Ed.), *G. Frege, Collected Papers on Mathematics, Logic, and Philosophy* (pp. 351–372). Blackwell.
French, S. (1999). Models and mathematics in physics. In J. Butterfield & C. Pagonis (Eds.), *From physics to philosophy* (pp. 187–207). Cambridge University Press.
French, S. (2014). *The structure of the world: Metaphysics and representation*. Oxford University Press.
French, S. (2015). Identity and individuality in quantum Physicis. In *Stanford Encyclopedia of philosophy*. E. Zalta (ed.). https://plato.stanford.edu/entries/qt-idind/#PII
French, S., & Krause, D. (2006). *Identity in physics: A historical, philosophical, and formal analysis*. Oxford University Press.
French, S., & Ladyman, J. (2003). Remodelling structural realism: Quantum physics and the metaphysics of structure. *Synthese, 136*, 31–56.

French, S., & Redhead, M. (1988). Quantum physics and the identity of indiscernibles. *The British Journal for the Philosophy of Science, 39*, 233–246.
Friedman, M. (2002). Physics, philosophy, and the foundations of geometry. *Dialogos, 79*, 121–143.
Galavotti, M. C., Dieks, D., Gonzalez, W., Hartmann, S., & Uebel, T. (Eds.). (2014). *New directions in the philosophy of science*. Springer.
Gale, R. (1966). McTaggart's analysis of time. *American Philosophical Quarterly, 3*, 145–152.
Gale, R. (1968). *The language of time*. Humanities Press.
Gallois, A. (1990). Occasional identity. *Philosophical Studies, 58*, 203–224.
Garcia-Carpintero, M. (2008). Relativism, vagueness, and what is said. In M. Garcia-Carpintero & M. Kölbel (Eds.), *Relative truth*. Oxford University Press.
Gaskin, R. (1995). *The sea Battle and the master argument: Aristotle and Diodorus Cronus on the metaphysics of the future*. De Gruyter.
Geach, P. (1972). Some problems about time. Reprinted in *Logic Matters*. P. Geach (ed.). University of California Press, pp. 302–318.
Geach, P. (1973). The future. *New Blackfriars, 54*, 208–218.
Geach, P. (1980). *Reference and generality: An examination of some medieval and modern theories* (3rd ed.). Cornell University Press.
Gell-Mann, M., & Hartle, J. (1993). Classical equations for quantum systems. *Physical Review D, 47*, 3345–3382.
Gettier, E. (1963). Is justified true belief knowledge? *Analysis, 23*, 121–123.
Gibbard, A. (1975). Contingent identity. *Journal of Philosophical Logic, 4*, 187–221.
Gilmore, C., Costa, D., & Calosi, C. (2016). Relativity and three four-Dimensionalisms. *Philosophy Compass, 11*, 102–120.
Gödel, K. (1949a). An example of a new type of cosmological solutions of Einstein's field equations of gravitation. *Institute for Advanced Study, 21*, 447–450.
Gödel, K. (1949b). A remark about the relationship between relativity theory and idealistic philosophy. In P. A. Schilpp (Ed.), *Albert Einstein – Philosopher-Scientist* (Vol. II, pp. 557–562). Open Court.
Godfrey-Smith, W. (1979). Special relativity and the present. *Philosophical Studies, 36*, 233–244.
Grandjean, V. (2020). Une Asymétrie temporelle : passé fermé et futur ouvert. In A. Declos & J.-B. Guillon (Eds.), *Les Principes Métaphysiques*. Ed. du Collège de France.
Grandjean, V. (2021a). How is the asymmetry between the open future and the fixed past to be characterized? *Synthese, 198*, 1863–1886.
Grandjean, V. (2021b). Symmetric and asymmetric theories of time. *Synthese, 199*, 14403–14426.
Grandjean, V. (2021c). Les Particuliers nus à la rescousse de la théorie du bloc en croissance'. In C. Tiercelin & A. Declos (dir.), *La Métaphysique du temps*. Ed. du Collège de France.
Grandjean, V. (2022). The bare past. *Philosophia*. https://doi.org/10.1007/s11406-022-00534-8
Grandjean, V., & Pascucci, M. (2021). The machine scenario: A computational perspective on alternative representations of indeterminism. *Minds and Machines, 31*, 59–74.
Greenough, P. (2008). Indeterminate truth. *Midwest Studies in Philosophy, 32*, 213–241.
Grey, W. (1997). Time and becoming. *Cogito, 11*, 215–220.
Grice, P. H. (1989). *Studies in the way of words*. Harvard University Press.
Groneberg, M. (2003). *La vérité du futur contingent : Łukasiewicz, Tarski ou van Fraassen ?* Actes du Colloque de la SOPHA.
Grünbaum, A. (1967). *Modern science and Zeno's paradoxes*. George Allen & Unwin.
Haack, S. (1974). *Deviant logic, fuzzy logic: Beyond the formalism*. Cambridge University Press.
Hacking, I. (1975). The identity of Indiscernibles. *Journal of Philosophy, 72*, 249–256.
Hagar, A. (2008). Kant and non-Euclidean geometry. *Kant-Studien, 99*, 81–82.
Hales, S. (2000). The problem of intuition. *American Philosophical Quarterly, 37*, 135–147.
Hare, C. (2009). *On myself, and other, less important subjects*. Princeton University Press.
Hare, C. (2013). Time – The emotional asymmetry. In H. Dyke & A. Bardon (Eds.), *A companion to the philosophy of time* (pp. 507–520). Wiley-Blackwell.

Haslanger, S. (2003). Persistence through time. In M. Loux & D. Zimmerman (Eds.), *The Oxford handbook of metaphysics* (pp. 315–354). Oxford University Press.

Hasle, P. (2012). The problem of predestination: As a prelude to A. N. Prior's tense logic. *Synthese, 188*, 331–347.

Hasle, P., & Øhrstrøm, P. (2004). The flow of time into logic – And computer science. *Bulletin of the European Association for Theoretical Computer Science, 82*, 191–226.

Hatfield, G. (2007). Kant on the perception of space (and time). In P. Guyer (Ed.), *The Cambridge companion to Kant and modern philosophy* (pp. 61–93). Cambridge University Press.

Hawking, S., & Penrose, R. (1996). *The nature of space and time*. Princeton University Press.

Hawking, S., King, A. R., & McCarthy, P. J. (1976). A new topology for curved Spacetime which incorporates the causal, differential, and conformal structure. *Journal of Mathematical Physics, 17*, 174–181.

Heathwood, C. (2005). The real Price of the dead past. *Analysis, 65*, 249–251.

Heathwood, C. (2007). On what will be: Reply to Westphal. *Erkenntnis, 67*, 137–142.

Heck, R. (2006). MacFarlane on relative truth. *Philosophical Issues, 16*, 88–100.

Heil, J. (2003). *From an ontological point of view*. Oxford University Press.

Heller, M. (1990). *The ontology of physical objects*. Cambridge University Press.

Henson, J. (2009). The causal set approach to quantum gravity. In D. Oriti (Ed.), *Approaches to quantum gravity: Toward a new understanding of space, time and matter* (pp. 323–345). Cambridge University Press.

Hinchliff, M. (2000). A defense of relativism in a relativistic setting. *Philosophy of Science, 67*, 575–586.

Hintikka, J. (1964). The once and Future Sea fight: Aristotle's discussion of future contingents in De Interpretatione IX. *The Philosophical Review, 73*, 461–492.

Hirsch, E. (2002). Quantifier variance and realism. *Noûs, 36*, 51–73.

Hirsch, E. (2006). Rashi's view on the open future: Indeterminateness and bivalence. In D. Zimmerman (Ed.), *Oxford studies in metaphysics* (Vol. 2, pp. 111–135). Clarendon press.

Holliday, W. (2012). Freedom and the fixity of the past. *Philosophical Review, 121*, 179–207.

Honderich, T. (1988). *A theory of determinism*. Oxford University Press.

Honderich, T. (2002). *How free are you?* (2nd ed.). Oxford University Press.

Horwich, P. (1987). *Asymmetries in time: Problems in the philosophy of science*. MIT Press.

Huggett, N. (Ed.). (1999a). *Space from Zeno to Einstein*. Harvard University Press.

Huggett, N. (1999b). What are quanta, and why does it matter? In *Proceedings of the 1994 Biennial Meeting of the Philosophy of Science Association* (PSA 1994, Vol. 2) (pp. 69–76). Philosophy of Science Association.

Hume, D. ([1748] 1975). *Enquiries concerning human understanding and concerning the principles of morals*. L. A. Selby-Bigge (Ed.). Clarendon Press.

Husserl, E. (1964). *The phenomenology of internal time-consciousness*. J. S. Churchill (Trans.). Martinus Nijhoff.

Iacona, A. (2007). Future contingents and Aristotle's fantasy. *Critica, Revista Hispanoamericana de Filosofia, 39*, 45–60.

IASP. (1994). Pain terms: A current list with definitions and notes on usage. In H. Merskey & N. Bogduk (Eds.), *Classification of chronic pain* (2nd ed., pp. 209–214). IASP Press.

Ingram, D. (2016). The virtues of Thisness presentism. *Philosophical Studies, 173*, 2867–2888.

Ingram, D. (2018). *Thisness presentism: An essay on time, Truth, and ontology*. Routledge.

Ingram, D. & Tallant, J. (2018). Presentism. In *Stanford Encyclopedia of Philosophy*. E. Zalta (ed.). https://plato.stanford.edu/entries/presentism/

Scott, M. (1995). Time and change. *The Philosophical. Quarterly, 45*, 213–218.

Irwin, T. (1992). Who discovered the Will? *Philosophical Perspectives, 6*, 453–473.

Ishiguro, H. (1976). Leibniz's theory of the Ideality of relations. In H. G. Frankfurt (Ed.), *Leibniz: A collection of critical essays* (pp. 191–213). Notre Dame University Press.

Ismael, J. (2003). Closed causal loops and the bilking argument. *Synthese, 136*, 305–320.

Ismael, J. (2013). Decision and the open future. In A. Bardon (Ed.), *The future of the philosophy of time* (pp. 149–168). Routledge.

Ismael, J. (2016). *How physics makes us Free*. Oxford University Press.

Jackson, F. (1994). Metaphysics by possible cases. *The Monist, 77*, 93–110.

Jackson, F. (1998). *From metaphysics to ethics: A defence of conceptual analysis*. Oxford University Press.

Kane, R. (2007). Libertarianism. In J. M. Fischer, R. Kane, D. Pereboom, & M. Vargas (Eds.), *Four views on free will* (pp. 5–43). Blackwell.

Kant, I. ([1787] 1998). *Critique of pure reason*. P. Guyer and A. Wood (Trans.). Cambridge University Press.

Kaplan, D. (1989). Demonstratives: An essay on the semantics, logic, metaphysics, and epistemology of demonstratives and other Indexicals. In J. Almog, J. Perry, & H. Wettstein (Eds.), *Themes from Kaplan* (pp. 481–566). Oxford University Press.

Keefe, R. (2000). *Theories of vagueness*. Cambridge University Press.

Keller, S. (2004). Presentism and truthmaking. In D. Zimmerman (Ed.), *Oxford studies in metaphysics* (Vol. 1, pp. 83–104). Oxford University Press.

Ketland, J. (2011). Identity and discernibility. *The Review of Symbolic Logic, 4*, 171–185.

Kiefer, C. (2011). Time in quantum gravity. In C. Callender (Ed.), *The Oxford handbook of philosophy of time* (pp. 663–678). Oxford University Press.

Kim, J. (1990). Supervenience as a philosophical concept. *Metaphilosophy, 21*, 1–27.

Kneale, W., & Kneale, M. (1962). *The development of logic*. Oxford University Press.

Knuuttila, S. (2015). Medieval theories of future contingents. In E. Zalta (Ed.), *Stanford Encyclopedia of philosophy*. https://plato.stanford.edu/entries/medieval-futcont/

Kölbel, M. (2003). Faultless Disagreement. *Proceedings of the Aristotelian Society., 104*, 53–73.

Koslicki, K. (2018). *Form, matter, substance*. Oxford University Press.

Kripke, S. ([1958] 2011). Letters to a. N. Prior. Published in T. Ploug & P. Øhrstrøm 'Branching time, indeterminism and tense logic'. *Synthese, 188*, 367–379.

Kripke, S. (1980). *Naming and necessity*. Blackwell.

Kronheimer, E. H., & Penrose, R. (1967). On the structure of causal spaces. *Mathematical Proceedings of the Cambridge Philosophical Society, 63*, 481–501.

Kutach, D. (2002). The entropy theory of counterfactuals. *Philosophy of Science, 69*, 82–104.

Kutach, D. (2007). The physical foundations of causation'. In H. Price & R. Corry (Eds.), *Causation, physics, and the constitution of reality: Russell's republic revisited* (pp. 327–350). Oxford University Press.

Kutach, D. (2011). The asymmetry influence. In C. Callender (Ed.), *The Oxford handbook of philosophy of time* (pp. 247–275). Oxford University Press.

Kutach, D. (2013). Time travel and time machines. In H. Dyke & A. Bardon (Eds.), *A companion to the philosophy of time* (pp. 301–314). Wiley-Blackwell.

Lacey, A. R. (1999). *Bergson: The arguments of the philosophers*. Routledge.

Ladyman, J. (1998). What is structural realism? *Studies in History and Philosophy of Science, 29*, 409–424.

Ladyman, J. (2005). Mathematical structuralism and the identity of Indiscernibles. *Analysis, 65*, 218–221.

Ladyman, J., & Ross, D. (2007). *Everything must go: Metaphysics naturalized*. Oxford University Press.

Ladyman, J., & Ross, D. (2013). The world in the data. In J. Ladyman, D. Ross, & H. Kincaid (Eds.), *Scientific metaphysics* (pp. 108–150). Oxford University Press.

Laplace, P.-S. ([1814] 1951). *A philosophical essay of probabilities*. F. Truscott and F. Emory (Trans.). Dover.

Latham, A. J., Miller, K., & Norton, J. (2020a). Do the folk represent time as essentially dynamical? *Inquiry*. https://doi.org/10.1080/0020174X.2020.1827027

Latham, A. J., Miller, K., & Norton, J. (2020b). An empirical investigation of purported passage phenomenology. *The Journal of Philosophy, 117*, 353–386.

Latham, A. J., Miller, K., & Norton, J. (2021a). Is our Naïve theory of time dynamical? *Synthese, 198*, 4251–4271.
Latham, A. J., Miller, K., & Norton, J. (2021b). An empirical investigation of the role of direction in our concept of time. *Acta Analytica, 36*, 25–47.
Latour, B. (2000). On the partial existence of existing and nonexisting objects. In L. Daston (Ed.), *Biographies of scientific objects*. University of Chicago Press.
Le Poidevin, R. (1991). *Change, cause, and contradiction*. Macmillan.
Le Poidevin, R. (2007). *The images of time, an essay on temporal representation*. Oxford University Press.
Le Poidevin, R. (2010). Time without change (in three steps). *American Philosophical Quarterly, 47*, 171–180.
Le Poidevin, R. (2013). Time and freedom. In H. Dyke & A. Bardon (Eds.), *A companion to the philosophy of time* (pp. 535–548). Wiley-Blackwell.
Le Poidevin, R., & Macbeath, M. (Eds.). (1993). *The philosophy of time*. Oxford University Press.
Leibniz, G. W. ([1765] 1981). New essays on human understanding. P. Remnant & J. Bennett (Trans.). Cambridge University Press.
Leitgeb, H., & Ladyman, J. (2008). Discussion note: Criteria of identity Structuralist ontology. *Philosophia Mathematica, 16*, 388–396.
Lewis, D. (1970). How to define theoretical terms. *Journal of Philosophy, 67*, 427–446.
Lewis, D. (1972). Psychophysical and theoretical identifications. *Australasian Journal of Philosophy, 50*, 249–258.
Lewis, D. (1973a). *Counterfactuals*. Blackwell.
Lewis, D. (1973b). Counterfactuals and comparative possibility. *Journal of Philosophical Logic, 2*, 418–446.
Lewis, D. (1976). The paradoxes of time travel. *American Philosophical Quarterly, 13*, 145–152.
Lewis, D. (1979a). Counterfactual dependence and Time's arrow. *Noûs, 13*, 455–476.
Lewis, D. (1979b). Attitudes De Dicto and De se. *The Philosophical Review, 88*, 513–543.
Lewis, D. (1982). Logic for equivocators. *Noûs, 16*, 128–130.
Lewis, D. (1983). New work for a theory of universals. *Australasian Journal of Philosophy, 61*, 343–377.
Lewis, D. (1986). *On the plurality of worlds*. Blackwell.
Lewis, D. (1993). Many, but almost one. In J. Bacon (Ed.), *Ontology, causality, and mind: Essays in honour of David Armstrong*. Cambridge University Press.
Lewis, D. (2001). Are we Free to break the Laws? *Theoria, 47*, 113–121.
Lewis, D. (2002). Tensing the copula. *Mind, 111*, 1–14.
Linsky, B., & Zalta, E. (1996). In Defense of the contingently nonconcrete. *Philosophical Studies, 8*, 283–294.
Lipman, M. (2015). On fine's fragmentalism. *Philosophical Studies, 172*, 3119–3133.
Locke, J. ([1689] 1975). In P. Nidditch (Ed.), *An essay concerning human understanding*. Clarendon Press.
Lockwood, M. (2005). *The labyrinth of time*. Oxford University Press.
Loewer, B. (2007). Counterfactuals and the second law. In H. Price & R. Corry (Eds.), *Causation, physics, and the constitution of reality: Russell's republic revisited* (pp. 293–326). Oxford University Press.
Lombard, L. (1999). On the alleged incompatibility of presentism and temporal parts. *Philosophia, 27*, 253–260.
Lorentz, H., Einstein, A., Minkowski, H., Weyl, H. & Sommerfeld, A. (1952). The principle of relativity: A collection of original memoirs on the special and general theory of relativity. W. Perrett & G. B. Jeffery (Trans.). Dover.
Loss, R. (2015). How to change the past in one-dimensional time. *Pacific Philosophical Quarterly, 96*, 1–11.
Loux, M. (1970). *Universals and particulars: Reading in ontology*. Doubleday.
Loux, M. (1998). *Metaphysics: A contemporary introduction*. Routledge.

Loux, M., & Zimmerman, D. (2003). *The Oxford handbook of metaphysics*. Oxford University Press.
Lovecraft, H. P. ([1927] 1974). *Supernatural horror in literature*. Dover Publications
Lowe, E. J. (1998). *The possibility of metaphysics*. Clarendon Press.
Lucas, J. R. (1986). The open future. In R. Flood & M. Lockwood (Eds.), *The nature of time* (pp. 125–134). Basil Blackwell.
Lucas, J. R. (1989). *The future*. Blackwell.
Lucas, J. R. (1999). A century of time. In J. Butterfield (Ed.), *The arguments of time* (pp. 1–20). Oxford University Press.
Ludlow, P. (2004). Presentism, triviality, and the varieties of Tensism. In D. Zimmerman (Ed.), *Oxford studies in metaphysics* (Vol. 1, pp. 21–36). Oxford University Press.
Łukasiewicz, J. (1930). Philosophische Bemerkungen zu Mehrwertigen Systemen des Aussagenkalküls. *Comptes rendus des séances de la Société des Sciences et des Lettres de Varsovie., 23*, 51–57.
Łukasiewicz, J. (1967). On determinism. In S. McCall (Ed.), *Polish logic 1920-1939* (pp. 19–39). Clarendon Press.
Luper-Foy, S. (1987). Annihilation. *The Philosophical Quarterly, 37*, 233–252.
Macfarlane, J. (2003). Future contingents and Relative Truth. *The Philosophical Quarterly, 53*, 322–336.
Macfarlane, J. (2005). Making sense of relative truth. *Proceedings of the Aristotelian Society, 105*, 321–339.
Macfarlane, J. (2007). Relativism and disagreement. *Philosophical Studies, 132*, 17–31.
Macfarlane, J. (2008). Truth in the garden of forking paths. In M. Garcia-Carpintero & M. Kölbel (Eds.), *Relative truth* (pp. 81–102). Oxford University Press.
Macfarlane, J. (2014). *Assessment sensitivity: Relative Truth and its applications*. Oxford University Press.
Machina, K. (1976). Truth, belief, and vagueness. In R. Keefe & P. Smith (Eds.), *Vagueness: A reader* (pp. 174–203). MIT Press.
Mackie, J. L. (1974). *The cement of the universe*. Oxford University Press.
Mackie, P. (2014). Counterfactuals and the fixity of the past. *Philosophical Studies, 168*, 397–415.
Maddy, P. (1990). *Realism in mathematics*. Oxford University Press.
Malament, D. (1977). The class of continuous Timelike curves determines the topology of Spacetime. *Journal of Mathematical Physics, 18*, 1399–1404.
Marcus, R. B. (1961). Modalities ad Intensional languages. *Synthese, 13*, 303–322.
Marcus, R. B. (1962). Interpreting Quantification. *Inquiry, 5*, 252–259.
Markosian, N. (1995). The open past. *Philosophical Studies, 79*, 95–105.
Markosian, N. (2004). A defense of presentism. In D. Zimmerman (Ed.), *Oxford studies in metaphysics* (Vol. 1, pp. 47–82). Oxford University Press.
Martinez, M. (2011). Travelling in branching time. *Disputatio, 4*, 59–75.
Marton, F., & Fensham, P. (1993). A Nobel's eye view of scientific intuition: Discussions with the Nobel prize-winners in physics, chemistry and Medecine. *International Journal of Science Education, 16*, 457–473.
Matthen, M. (2009). Chicken, eggs, and speciation. *Noûs, 43*, 94–115.
Maudlin, T. (2002). Remarks on the passing of time. *Proceedings the Aristotelian Society, 102*, 259–274.
Maudlin, T. (2007). *The metaphysics within physics*. Oxford University Press.
Maudlin, T. (2011). *Quantum non-locality and relativity*. Wiley-Blackwell.
Maudlin, T. (2012). *Philosophy of physics: Space and time*. Princeton University Press.
Maudlin, T. (2019). *Philosophy of physics: Quantum theory*. Princeton University Press.
Maxwell, N. (1985). Are Probabilism and special relativity incompatible? *Philosophy of Science, 52*, 23–43.
Maxwell, C. ([1877] 2010). *Matter and motion*. Cambridge University Press.
Mayr, E. (1942). *Systematics and the origin of species*. Columbia University Press.

Mays, W. (1972). Whithead and the philosophy of time. In J. T. Fraser, F. C. Haber, & G. H. Müller (Eds.), *The study of time* (pp. 354–369). Springer.
McCall, S. (1976). Objective time flow. *Philosophy of Science, 43*, 337–362.
McCall, S. (1994). *A model of the universe*. Clarendon Press.
McDaniel, K. (2003). No paradox of multi-location. *Analysis, 63*, 309–311.
McDaniel, B. (2010). Presentism and absence causation: An exercise in mimicry. *Australasian Journal of Philosophy, 88*, 323–332.
McKenna, M., & Coates, J. (2015). Compatibilism. In E. Zalta (Ed.), *Stanford Encyclopedia of Philosophy*. https://plato.stanford.edu/entries/compatibilism/
McPherran, M. (1988). Plato's particulars. *Southern Journal of Philosophy, 26*, 527–553.
McTaggart, J. (1908). The unreality of time. *Mind, 17*, 457–474.
McTaggart, J. ([1896] 2011). *Studies in the Hegelian dialectic*. Cambridge University Press.
Mellor, D. H. (1974). Special relativity and the present truth. *Analysis, 34*, 74–78.
Mellor, D. H. (1981). *Real time*. Cambridge University Press.
Mellor, D. H. (1982). Theoretically structured time. *Philosophical Books, 23*, 65–69.
Mellor, D. H. (1987). Fixed past, unfixed future. In B. M. Taylor (Ed.), *Michael Dummett: Contributions to philosophy* (pp. 166–186). Springer.
Mellor, D. H. (1998). *Real time II*. Routledge.
Merleau-Ponty, M. (1962). *Phenomenology of perception*. C. Smith (Trans.). Routledge and Kegan Paul.
Merricks, T. (1995). On the incompatibility of enduring and Perduring entities. *Mind, 104*, 523–531.
Merricks, T. (1999). Persistence, parts, and presentism. *Noûs, 33*, 421–438.
Merricks, T. (2006). Goodbye growing block. In D. Zimmerman (Ed.), *Oxford studies in metaphysics* (Vol. 2). Oxford University Press.
Mertz, D. W. (2001). Individuation and instance ontology. *Australasian Journal of Philosophy, 79*, 45–61.
Meyer, U. (2005). The Presentist's dilemma. *Philosophical Studies, 122*, 213–225.
Meyer, U. (2011). The triviality of presentism. In R. Ciuni, K. Miller, & G. Torrengo (Eds.), *New papers on the present: Focus on presentism*. Philosophia Verlag.
Meyer, U. (2012). Explaining Causal Loops. *Analysis, 72*, 259–264.
Meyer, U. (2013). *The nature of time*. Oxford University Press.
Meyer, U. (2016). Consciousness and the present. In Y. Dolev & M. Roubach (Eds.), *Cosmological and psychological time* (pp. 143–153). Springer.
Meyer, U. (2019). A review of Fabrice Correia and Sven Rosenkranz *nothing to come: A defence of the growing block theory of time*. Notre Dame Philosophical Reviews. https://ndpr.nd.edu/news/nothing-to-come-a-defence-of-the-growing-block-theory-of-time/
Miller, K. (2005). Time travel and the open future. *Disputatio, 1*, 223–232.
Miller, K. (2013). Presentism, Eternalism, and the Growing Block. In H. Dyke & A. Bardon (Eds.), *A companion to the philosophy of time* (pp. 345–364). Wiley-Blackwell.
Miller, K. (2019). The cresting wave: A new moving spotlight theory. *Canadian Journal of Philosophy, 49*, 94–122.
Minkowski, H. ([1908] 1923). The Union of space and time. In A. Sommerfeld (Ed.), W. Perrett & G. B. Jeffery (TRANS.), *The principle of relativity* (pp. 75–80). Dover.
Misner, C., Thorne, K., & Wheeler, J. A. (1973). *Gravitation*. W. H. Freeman and Company.
Monton, B. (2006). Presentism and quantum gravity. In D. Dieks (Ed.), *The ontology of Spacetime* (pp. 263–280). Kluwer.
Moreland, J. P. (1998). Theories of individuation: A reconsideration of bare particulars. *Pacific Philosophical Quarterly, 79*, 251–263.
Moreland, J. P. (2001). *Universals*. McGill-Queen's University Press.
Morganti, M. (2004). On the preferability of epistemic structural realism. *Synthese, 142*, 81–107.
Morganti, M. (2011). Substrata and properties: From bare particulars to Supersubstantivalism? *Metaphysica, 12*, 183–195.

Morganti, M., & Takho, T. (2017). Moderately naturalistic metaphysics. *Synthese, 194*, 2557–2580.

Morris, M., Nisbett, R., & Peng, K. (1995). Causal attribution across domains and cultures. In D. Sperber, D. Premack, & A. J. Premack (Eds.), *Causal cognition* (pp. 577–612). Oxford University Press.

Müller, T. (2007). Branch dependence in the "consistent histories" approach to quantum mechanics. *Foundations of Physics, 37*, 253–276.

Müller, T. (2012). Branching in the landscape of possibilities. *Synthese, 188*, 41–65.

Müller, T. (Ed.). (2014). *Nuel Belnap on indeterminism and Free action*. Springer.

Müller, T., Nuel, B., & Kishida, K. (2008). Funny business in branching space-times: Infinite modal correlations. *Synthese, 164*, 141–159.

Myro, G. (1985). Identity and time. In R. Grandy & R. Warner (Eds.), *Philosophical grounds of rationality: Intentions, Categories, Ends* (pp. 383–410). Oxford University Press.

Nagel, T. (1979). Death. In T. Nagel (Ed.), *Mortal questions*. Cambridge University Press.

Nerlich, G. (1995). Falling branches and the flow of time. *Australasian Journal of Philosophy, 76*, 309–316.

Neurath, O. (1973). Empiricism and sociology. M. Neurath & R. S. Cohen (eds.). D. Reidel Publishing Company.

Newton, I. (1729). *The mathematical principles of natural philosophy*. A. Motte (Trans.). Printed for Benjamin motte, at the *middle-Temple-gate. Fleetstreet.*

Nishimura, H. (1979). Is the semantics of branching structures adequate for non-metric Ockhamist tense logics. *Journal of Philosophical Logic, 8*, 477–478.

Norton, J. (1993). General covariance and the foundations of general relativity: Eight decades of dispute. *Reports on Progress in Physics, 56*, 791–858.

Norton, J. (2010). Time really passes. *Humana.Mente: Journal of Philosophical Studies., 13*, 23–34.

Norton, J. (2014). What can we learn about the ontology of space and time from the theory of relativity. In L. Sklar (Ed.), *Physical theory: Method and interpretation*. Oxford University Press.

Norton, J. (2015). The burning fuse model of unbecoming time. *Studies in History and Philosophy of Science Part B: Studies in History and Philosophy of Modern Physics, 52*, 103–105.

Norton, J. (2018). If time travel to our location is possible, we do not live in a branching universe. *Analysis, 78*, 260–266.

O'leary-Hawthorne, J. (1995). The bundle theory of substance and the identity of Indiscernibles. *Analysis, 55*, 191–196.

O'leary-Hawthorne, J., & Cover, J. A. Framing the thisness issue. *Australasian Journal of Philosophy, 75*, 102–108.

Oaklander, L. N. (2001). Is there a difference between the metaphysics of A- and B-time? *Journal of Philosophical Research, 26*, 23–36.

Ockham, W. (1983). *Predestination, God's foreknowledge, and future contingents*, 2nd edn. M. McCord Adams & N. Kretzmann (Trans.). Hackett Publishing Company.

Øhrstrøm, P. (1981). Problems regarding the future operator in an Indeterministic tense logic. *Danish Yearbook of Philosophy, 18*, 81–95.

Øhrstrøm, P. (1983). Richard Lavenham on future contingents. *Cahiers de l'Institut du Moyen-Âge Grec et Latin., 44*, 180–186.

Øhrstrøm, P. (1984). Anselm, Ockham and Leibniz on divine foreknowledge and human freedom. *Erkenntnis, 21*, 209–222.

Øhrstrøm, P. (2014). What William of Ockham and Luis de Molina would have said to Nuel Belnap: A discussion of some arguments against the thin red line. In T. Müller (Ed.), *Nuel Belnap on indeterminism and free action* (pp. 175–190).

Øhrstrøm, P., & Hasle, P. (1995). *Temporal logic: From ancient ideas to artificial intelligence*. Kluwer Academic Publishers.

Øhrstrøm, P., & Hasle, P. (2020). Future Contingents. In E. Zalta (Ed.), *Stanford Encyclopedia of Philosophy*. https://plato.stanford.edu/entries/future-contingents/

Palter, R. M. (1960). *Whitehead's philosophy of science*. University of Chicago Press.

Papineau, D. (1990). Why supervenience? *Analysis, 50*, 66–71.

Pappas, G. (2014). Internalist vs. externalist conceptions of epistemic justification. In E. Zalta (Ed.), *Stanford Encyclopedia of philosophy*. https://plato.stanford.edu/entries/justep-intext/

Parsons, T. (1994). Ruth Barcan Marcus and the Barcan formula. In W. Sinnott-Armstrong, D. Raffman, & N. Asher (Eds.), *Modality, morality and belief: Essays in honor of Ruth Barcan Marcus*. Cambridge University Press.

Parsons, J. (2002). A-theory for B-theorists. *The Philosophical Quarterly, 52*, 1–20.

Paul, L. A. (2002). Logical parts. *Noûs, 36*, 578–596.

Pereboom, D. (2005). Defending hard Incompatibilism. *Midwest Studies in Philosophy, 29*, 228–247.

Perry, J. (2004). Compatibilist options. In J. K. Campbell, M. O'Rourke, & D. Shier (Eds.), *Freedom and determinism* (pp. 231–254). MIT Press.

Petkov, V. (2006). Is there an alternative to the block universe view? In D. Dieks (Ed.), *The ontology of Spacetime* (1st ed., pp. 207–228).

Petkov, V. (Ed.). (2010). *Space, time, and Spacetime: Physical and philosophical implications of Minkowski's unification of space and time*. Springer.

Pike, N. (1965). Divine omniscience and voluntary action. *The Philosophical Review, 74*, 27–46.

Pike, N. (1966). Of god and freedom: A rejoinder. *The Philosophical Review, 75*, 369–379.

Placek, T. (2000). Stochastic outcomes in branching space-time: Analysis of Bell's theorem. *The British Journal for the Philosophy of Science, 51*, 445–475.

Plantinga, A. (1974). *The nature of necessity*. Oxford University Press.

Plantinga, A. (1976). Actualism and possible worlds. *Theoria, 42*, 139–160.

Plantinga, A. (1983). On Existentialism. *Philosophical Studies, 44*, 1–20.

Plantinga, A. (1986). On Ockham's way out. *Faith and Philosophy, 3*, 235–269.

Plato. (2000). *Timaeus*. D. J. Zeyl (Trans.). Hackett Publishing Co.

Ploug, T., & Øhrstrøm, P. (2012). Branching time, indeterminism and tense logic. *Synthese, 188*, 367–379.

Pooley, O. (2013). Relativity, the open future and the passage of time. *Proceedings of the Aristotelian Society, 113*, 321–363.

Popper, K. (1988). *The open universe: An argument for indeterminism*. Routledge.

Prawitz, D. (2009). Logical determinism and the principle of bivalence. In F. Stoutland (Ed.), *Philosophy Probings: Essays on Von Wright's later work* (pp. 11–35). Automatic Press.

Price, H. (1994). A neglected route to realism about quantum mechanics. *Mind, 103*, 303–336.

Price, H. (1996). *Time's arrow and Archimedes' point: New directions for the physics of time*. Oxford University Press.

Price, H. (2007). Causal perspectivalism. In H. Price & R. Corry (Eds.), *Causation, physics, and the constitution of reality: Russell's republic revisited* (pp. 250–292). Oxford University Press.

Price, H. (2011). The flow of time. In C. Callender (Ed.), *The Oxford handbook of philosophy of time* (pp. 276–311). Oxford University Press.

Prior, A. (1953). Three-valued logic and future contingents. *The Philosophical Quarterly, 3*, 317–326.

Prior, A. (1957). *Time and modality*. Clarendon Press.

Prior, A. (1959). Thank goodness that's over. *Philosophy, 34*, 12–17.

Prior, A. (1962). *Formal logic*. Clarendon Press.

Prior, A. (1966). Postulates for tense-logic. *American Philosophical Quarterly, 3*, 153–161.

Prior, A. (1967). *Past, present and future*. Clarendon Press.

Prior, A. (1970). The notion of the present. *Studium Generale, 23*, 245–248.

Prior, A. (1996). Some free thinking about time. In J. Copeland (Ed.), *Logic and reality* (pp. 47–51). Clarendon Press.

Prior, A. (2003). Changes in events and changes in things. In P. Hasle, P. Øhrstrøm, T. Braüner, & J. Copeland (Eds.), *Papers on time and tense* (pp. 7–19). Oxford University Press.

Prosser, S. (2000). A new problem for the A-theory of time. *The Philosophical Quarterly, 50*, 494–498.

Prosser, S. (2007). Could we experience the passage of time. *Ratio, 20*, 75–90.
Pruss, A. (2003). David Lewis's counterfactual arrow of time. *Noûs, 37*, 606–637.
Psillos, S. (2001). Is structural realism possible? *Philosophy of Science, 68*, 13–24.
Psillos, S. (2006). The structure, the whole and nothing but the structure? *Philosophy of Science, 73*, 560–570.
Putnam, H. (1967). Time and physical geometry. *Journal of Philosophy, 64*, 240–247.
Quine, W. V. O. (1960). *Word and object*. MIT Press.
Quine, W. V. O. (1966). *Selected logic papers*. Random House.
Quine, W. V. O. (1981). In W. V. O. Quine (Ed.), *Things and their place in theories. Theories and things* (pp. 1–23). Harvard University Press.
Rakic, N. (1997). Past, present, future and special relativity. *British Journal for the Philosophy of Science, 48*, 257–280.
Reichenbach, H. ([1956] 1971). *The direction of time*. University of California Press.
Reichenbach, H. (1957). *The philosophy of space and time*. Dover Books.
Reichenbach, H. (1969). *Axiomatization of the theory of relativity*. University of California Press.
Reimer, M. (2019). Reference. In E. Zalta (Ed.), *Stanford Encyclopedia of philosophy*. https://plato.stanford.edu/entries/reference/
Rescher, N., & Urquhart, A. (1971). *Temporal logic*. Springer.
Rettler, B., & Bailey, A. (2017). Object. In E. Zalta (Ed.), *Stanford Encyclopedia of Philosophy*. https://plato.stanford.edu/entries/object/
Rideout, D., & Sorkin, R. (2000). Classical sequential growth dynamics for causal sets. *Physical Review D, 61*, 024002-1-16.
Riemann, B. ([1854] 1953). Über die Hypothesen, welche der Geometrie zu Grunde liegen'. *The collected works of B. Riemann*. Dover.
Rietdijk, W. (1966). A rigorous proof of determinism derived from the special theory of relativity. *Philosophy of Science, 33*, 341–344.
Robb, A. (1936). *Geometry of time and space*. Cambridge University Press.
Robertson, H. P. (1949). Geometry as a branch of physics. In P. Schilpp (Ed.), *Albert Einstein: Philosopher-Scientist* (pp. 315–332). Library of Living Philosophers.
Rosen, G. (2017). Abstract objects. In E. Zalta (Ed.), *Stanford Encyclopedia of philosophy*. https://plato.stanford.edu/entries/abstract-objects/
Rosenkrantz, G. S. (1993). *Haecceity*. Kluwer Academic Publishers.
Rosenkranz, S. (2013). Determinism, the open future and branching time. In F. Correia & A. Iacona (Eds.), *Around the tree* (pp. 47–72).
Rovelli, C. (2018). *The order of time*. Penguin Books Ltd.
Rovelli, C. (2020). Space and time in loop quantum gravity. In N. Huggett, K. Matsubara, & C. Wüthrich (Eds.), *Beyond Spacetime: The foundations of quantum gravity* (pp. 143–153). Cambridge University Press.
Russell, B. (1905). On denoting. *Mind, 14*, 479–493.
Russell, B. (1911). Knowledge by acquaintance and knowledge by description. *Proceedings of the Aristotelian Society, 11*, 108–128.
Russell, B. (1940). *An inquiry into the meaning of truth*. Allen and Unwin.
Russell, B. (1957). Mr. Strawson on referring. *Mind, 66*, 385–389.
Russell, B. ([1903] 2010). *Principles of mathematics*. Routledge.
Russell, B., & Whitehead, A. N. (1957). *Principia Mathematica* (2nd ed.). Cambridge University Press.
Ryle, G. ([1953] 2015). It Was To Be. *Dilemmas: The Tarner Lectures* (pp. 13–30). Cambridge University Press.
Sainsbury, M. (1994). Why the World cannot be vague. *Southern Journal of Philosophy, 33*, 63–82.
Salmon, W. C. (Ed.). (1979). *Hans Reichenbach: Logical empiricist*. D. Reidel.
Sarkar, S., & Pfeifer, J. (Eds.). (2006). *The philosophy of science: An Encyclopedia* (Vol. 1). Routledge.
Sattig, T. (2014). Mereological indeterminacy: Metaphysical but not fundamental. In K. Akiba & A. Abasnezhad (Eds.), *Vague objects and vague identity: New essays on ontic vagueness*. Springer.

Saunders, S. (1995). Time, quantum mechanics and Decoherence. *Synthese, 102*, 235–266.
Saunders, S. (1996). Time, quantum mechanics and tense. *Synthese, 107*, 19–53.
Saunders, S. (1997). Time, quantum mechanics and probability. *Synthese, 114*, 373–404.
Saunders, S. (2002). How relativity contradicts presentism. *Royal Institute of Philosophy Supplements, 50*, 277–292.
Saunders, S. (2003). Physics and Leibniz's principles. In K. Brading & E. Castellani (Eds.), *Symmetries in physics: Philosophical reflections*. Cambridge University Press.
Savitt, S. (Ed.). (1995). *Time's arrows today: Recent physical and philosophical work on the direction of time*. Cambridge University Press.
Savitt, S. (2000). There's no time like the present (in Minkowski Spacetime). *Philosophy of Science, 67*(Proceedings), 563–574.
Savitt, S. (2006). Presentism and Eternalism in perspective. In D. Dieks & M. Redei (Eds.), *The ontology of Spacetime* (pp. 111–128). Elsevier.
Schaffer, J. (2009). Spacetime the one substance. *Philosophical Studies, 145*, 131–148.
Schmidt, M. (2008). On Spacetime, points, and bare particulars. *Metaphysica, 9*, 69–77.
Schrödinger, E. (1935). The present situation in quantum mechanics. In J. Wheeler & W. Zurek (Eds.), *Quantum theory and measurement* (pp. 152–167). Princeton University Press.
Schwitzgebel, E. (2019). Belief. In E. Zalta (Ed.), *Stanford Encyclopedia of philosophy*. https://plato.stanford.edu/entries/belief/
Sellars, W. (1962). Philosophy and the scientific image of man. In R. Colodny (Ed.), *Science, perception, and reality* (pp. 35–78). University of Pittsburgh Press.
Shapiro, S. (1997). *Philosophy of mathematics: Structure and ontology*. Oxford University Press.
Shoemaker, S. (1969). Time without change. *Journal of Philosophy, 66*, 363–381.
Shoemaker, S. (1980). Causality and properties. In P. van Inwagen (Ed.), *Time and cause* (pp. 109–135). D. Reidel.
Sider, T. (2001). *Four-dimensionalism*. Oxford University Press.
Sider, T. (2002). Time travel, coincidences and counterfactuals. *Philosophical Studies, 110*, 115–138.
Sider, T. (2005). Traveling in A- and B-time. *The Monist, 88*, 329–335.
Sider, T. (2006a). Quantifiers and temporal ontology. *Mind, 115*, 75–97.
Sider, T. (2006b). Bare Particulars. *Philosophical Perspectives, 20*, 387–397.
Sider, T. (2011). *Writing the book of the World*. Oxford University Press.
Simons, P. (2000). Continuants and occurrents. *The Aristotelian Society, 74*, 78–101.
Simons, P. (2003). Events. In M. Loux & D. Zimmerman (Eds.), *The Oxford handbook of metaphysics* (pp. 357–385). Oxford University Press.
Sklar, L. (1974). *Space, time, and Spacetime*. University of California Press.
Sklar, L. (1992). *Philosophy of physics*. Oxford University Press.
Sklar, L. (1993). *Physics and chance: Philosophical issues in the foundations of statistical mechanics*. Cambridge University Press.
Skow, B. (2005). *Once upon a space time*, PhD Thesis. New York University.
Skow, B. (2010). Deep metaphysical indeterminacy. *The Philosophical Quarterly, 60*, 851–858.
Skow, B. (2012). One second per second. *Philosophy and Phenomenological Research, 85*, 377–389.
Skow, B. (2015). *Objective becoming*. Oxford University Press.
Smart, J. (1949). The river of time. *Mind, 58*, 483–494.
Smart, J. J. C. (1963). *Philosophy and scientific realism*. Routledge & Kegan Paul.
Smart, J. J. C. (1987). Time and becoming. Reprinted in *Essays metaphysical and moral*. Smart, J. J. C. (ed.). Basil Blackwell, pp. 78–90.
Smeenk, C. (2013). Time in cosmology. In H. Dyke & A. Bardon (Eds.), *A companion to the philosophy of time* (pp. 201–219). Wiley-Blackwell.
Smeenk, C., & Wüthrich, C. (2011). Time travel and time machines. In C. Callender (Ed.), *The handbook of philosophy of time* (pp. 577–630). Oxford University Press.

Smith, Q. (1985). The mind-Independence of temporal becoming. *Philosophical Studies, 47*, 109–119.
Smith, Q. (2002). Time and degrees of existence: A theory of 'degree presentism'. In C. Callender (Ed.), *Time, reality and experience* (pp. 119–136). University Press.
Smith, N. (2011). Inconsistency in the A-theory. *Philosophical Studies, 156*, 231–247.
Smith, N. (2017). I'd do anything to change the past (but I can't do 'that'). *American Philosophical Quarterly., 54*, 153–168.
Sorensen, R. (2001). *Vagueness and contradiction*. Oxford University Press.
Sorkin, R. (2002). Causal sets: Discrete gravity. *Notes for the Valdivia Summer School,* Jan 2002. https://arxiv.org/abs/gr-qc/0309009
Sorkin, R. (2006). Geometry from order: Causal sets. *Einstein Online, 2*, 1007.
Sorkin, R. (2007). Relativity theory does not imply that the future already exists: A counterexample. In V. Petkov (Ed.), *Relativity and the dimensionality of the World* (pp. 153–161). Springer.
Sosa, E. (2009). A defense of the use of intuitions in philosophy. In M. Bishop & D. Murphy (Eds.), *Stich and his critics* (pp. 101–112). Blackwell.
Spinoza, B. (2002). *Complete works*. S. Shirley (Trans.), M. L. Morgan (ed.). Hackett Publishing.
Stapp, H. (1986). Einstein time and process time. In D. R. Griffin (Ed.), *Physics and the ultimate significance of time* (pp. 264–270). State University of New York Press.
Stein, H. (1968). On Einstein-Minkowski space-time. *Journal of Philosophy, 65*, 5–23.
Stein, H. (1991). On relativity theory and openness of the future. *Philosophy of Science, 58*, 147–167.
Stephens, C. (1989). The most reliable time. *Technology and Culture, 30*, 1–24.
Strang, C. (1960). Aristotle and the sea-Battle. *Mind, 69*, 447–465.
Strawson, P. (1950). On referring. *Mind, 59*, 320–344.
Strawson, P. ([1959] 2003). *Individuals*. Routledge.
Studtmann, P. (2013). Aristotle's categories. In E. Zalta (Ed.), *Stanford Encyclopedia of philosophy*. https://plato.stanford.edu/entries/aristotle-categories/
Sumner, L. W. (1976). A matter of life and death. *Noûs, 10*, 145–171.
Surya, S. (2019). The causal set approach to quantum gravity. *Living Reviews in Relativity, 22*, Article n°5.
Swinburne, R. (1979). *The existence of god*. Oxford University Press.
Swinburne, R. (1989). *Responsibility and atonement*. Clarendon Press.
Tallant, J. (2013). Intuitions in physics. *Synthese, 190*, 2959–2980.
Tallant, J. (2014). Metaphysics, intuitions and physics. *Ratio, 28*, 286–301.
Tallant, J. (2015). Metaphysics, intuitions and physics. *Ratio, 28*, 286–301.
Tarski, A. (1952). The semantic conception of truth. In L. Linsky (Ed.), *Semantics and the philosophy of language*. University of Illinois Press.
Thomas, J. L. H. (1991). Against the fantasts. *Philosophy, 66*, 349–367.
Thomason, R. (1969). Species, determinates and natural kinds. *Noûs, 3*, 95–101.
Thomason, R. (1970). Indeterminist time and truth value gaps. *Theoria, 36*, 264–281.
Thomason, R. (1981). Deontic logic as founded on tense logic. In R. Hilpinen (Ed.), *New studies in deontic logic* (pp. 165–176).
Thomason, R. (1984). Combinations of tense and modality. In D. Gabby & F. Guenthner (Eds.), *Handbook of philosophical logic. Vol. II: Extensions of classical logic*. D. Reidel Publishing Company.
Thomason, R., & Gupta, A. (1980). A theory of conditionals in the context of branching time. *The Philosophical Review, 89*, 65–90.
Thomson, J. (1976). Killing, letting die, and the trolley problem. *The Monist, 59*, 204–217.
Todd, P. (2016). Future contingents are all false! On behalf of a Russellian open future. *Mind, 125*, 775–798.
Tooley, M. (1997). *Time, tense and causation*. Clarendon Press.
Torre, S. (2011). The open future. *Philosophy Compass, 6*, 360–373.
Torrengo, G. (2008). *Time and cross-temporal relations*. Mimesis.
Tugby, M. (2021). Grounding theories of powers. *Synthese, 198*, 11187–11216.

Uffink, J. (2001). Bluff your way in the second law of thermodynamics. *Studies in the History and Philosophy of Modern Physics, 32*, 305–394.
Unger, P. (1979). There are no ordinary things. *Synthese, 4*, 117–154.
Van Cleve, J. (1985). Three versions of the bundle theory. *Philosophical Studies, 47*, 95–107.
Van Fraassen, B. (1966). Singular terms, truth-value gaps and free logic. *Journal of Philosophy, 63*, 481–495.
Van Fraassen, B. (1968). Presupposition, implication, and self-reference. *Journal of Philosophy, 65*, 136–152.
Van Inwagen, P. (1981). The doctrine of arbitrary Undetached parts. *Pacific Philosophical Quarterly, 62*, 123–137.
Van Inwagen, P. (1983). *An essay on free will*. Oxford University Press.
Van Inwagen, P. (1989). When is the will free? *Philosophical Perspectives, 3*, 399–342.
Van Inwagen, P. (1990). Four-dimensional objects. *Noûs, 24*, 245–255.
Varadarajan, M., & Rideout, D. (2006). A general solution for classical sequential growth dynamics of causal sets. *Physical Review D, 73*, 104021.
Vihvelin, K. (2003). Arguments for incompatibilism. In E. Zalta (Ed.), *Stanford Encyclopedia of philosophy*. https://plato.stanford.edu/entries/incompatibilism-arguments/
Von Wright, G. H. (1979). Time, truth, and necessity. In C. Diamond & J. Teichman (Eds.), *Intention and intentionality* (pp. 237–250). Cornell University Press.
Wallace, D. (2013). The arrow of time in physics. In H. Dyke & A. Bardon (Eds.), *A companion to the philosophy of time* (pp. 262–281). Wiley-Blackwell.
Warmbrod, K. (2017). Time, change and time without change. *Synthese, 194*, 3047–3067.
Wasserman, R. (2017). *The paradoxes of time travel*. Oxford University Press.
Waterlow, S. (1974). Backwards causation and continuing. *Mind, 83*, 372–387.
Wawer, J. (2018). Some problems with the Russellian open future. *Acta Analytica, 33*, 413–425.
Wegener, M. (Ed.). (1999). *Time, creation and world-order*. Aarhus University Press.
Weinert, F. (2004). *The scientist as philosopher: Philosophical consequences of great scientific discoveries*. Springer.
Weinert, F. (2013a). *The march of time*. Springer.
Weinert, F. (2013b). The past-future asymmetry. In R. Steineck & C. Clausius (Eds.), *Origins and future: Time inflected and reflected* (pp. 139–166). Brill.
Weinert, F. (2016). *The demons of science: What they can and cannot tell us about our world*. Springer.
Weinert, F. (2017). Tracing the arrows of time. In I. T. Durham & D. Rickles (Eds.), *Information and interaction: Eddington, wheeler, and the limits of knowledge* (pp. 73–101). Springer.
Weinert, F. (2019). Einstein and Kant. *Philosophy, 80*, 585–593.
Weyl, H. (1966). *Philosophy of mathematics and natural science*. O. Helmer (Trans.). Princeton University Press.
Whitehead, A. N. (1919). *An enquiry concerning the principles of natural knowledge*. Cambridge University Press.
Whitehead, A. N. (1979). *Process and reality: An essay in cosmology*. The Free Press.
Whitehead, A. N. ([1934] 2011). *Nature and life*. Chicago University Press. Reprinted Cambridge: Cambridge University Press
Wiggins, D. (1968). On being in the same place at the same time. *The Philosophical Review, 77*, 90–95.
Wildman, N. (2015). Load bare-ing particulars. *Philosophical Studies, 172*, 1419–1434.
Will, C. (2005). Was Einstein right? Testing relativity at the centenary. In A. Ashtekar (Ed.), *100 years of relativity; space-time structure: Einstein and beyond* (pp. 205–227). World Scientific.
Williams, D. C. (1951a). The myth of passage. *Journal of Philosophy, 48*, 457–472.
Williams, D. C. (1951b). The sea fight tomorrow. In P. Henle (Ed.), *Structure, meaning and method* (pp. 262–288). Liberal Arts Press.
Williams, C. (1996). The metaphysics of a and B-time. *The Philosophical Quarterly., 46*, 379–393.
Williams, C. (1998a). B-time transition. *Philosophical Inquiry, 20*, 59–63.
Williams, C. (1998b). A Bergsonian approach to A- and B-time. *Philosophia, 73*, 379–393.

Williams, R. (2008a). Multiple actualities and Ontically vague identity. *The Philosophical Quarterly, 52*, 134–154.

Williams, R. (2008b). Ontic vagueness and metaphysical indeterminacy. *Philosophy Compass, 3*, 763–788.

Williamson, T. (1994). *Vagueness*. Routledge.

Williamson, T. (2000). Existence and contingency. *Proceedings of the Aristotelian Society, 73*, 181–203.

Williamson, T. (2013). *Modal logic as metaphysics*. Oxford University Press.

Wilson, J. (2013). A determinable-based account of metaphysical indeterminacy. *Inquiry, 56*, 359–385.

Wilson, J. (2016). Are there indeterminate states of affairs? Yes'. In E. Barnes (Ed.), *Current controversies in metaphysics* (pp. 105–119). Routledge.

Winnie, J. (1970). Special relativity without one-way velocity assumptions. *Philosophy of Science, 37*, 81–99.

Wüthrich, C. (2005). To quantize or not to quantize: Fact and folklore in quantum gravity. *Philosophy of Science, 72*, 777–788.

Wüthrich, C. (2010). No presentism in quantum gravity. In V. Petkov (Ed.), *Space, time and Spacetime: Physical and philosophical implications of Minkowski's unification of space and time* (pp. 257–278). Springer.

Wüthrich, C. (2011). The fate of presentism in modern physics. In R. Ciuni, K. Miller, & G. Torrengo (Eds.), *New papers on the present – Focus on presentism*. Philosophia Verlag.

Wüthrich, C. (2013). The structure of causal sets. *Journal for General Philosophy of Science, 43*, 223–241.

Wüthrich, C., & Callender, C. (2016). What becomes of a causal set? *The British Journal for the Philosophy of Science, 0*, 1–19.

Zagzebski, L. (2017). Foreknowledge and free Will. In *Stanford Encyclopedia of philosophy*, E. Zalta (ed.). https://plato.stanford.edu/entries/free-will-foreknowledge/

Zanardo, A. (1990). Axiomatization of Peircean branching-time logic. *Studia Logica, 49*, 183–195.

Zanardo, A. (2003). First-order and second-order aspects of branching-time semantics. In *Preprint Dipartimento di Matematica Università di Padova. N° 3*. Presented at HPLMC-02, Second International Workshop of the History and Philosophy of Logic, Mathematics and Computation (Donostia-San Sebastian, Spain, 7–9 November, 2002).

Zanardo, A. (2006). Moment/history in Prior's logics of branching-time. *Synthese, 150*, 483–507.

Zeeman, E. C. (1964). Causality implies the Lorentz group. *Journal of Mathematical Physics, 5*, 490–493.

Zimmerman, D. (1996). Persistence and presentism. *Philosophical Papers, 25*, 115–126.

Zimmerman, D. (1997). Chisholm and the essences of events. In L. E. Hahn (Ed.), *The philosophy of Roderick M. Chisholm* (pp. 73–100). Open Court.

Zimmerman, D. (1998). Temporary Intrinsics and presentism. In P. Van Inwagen & D. Zimmerman (Eds.), *Metaphysics: The big questions* (pp. 206–219). Blackwell.

Zimmerman, D. (Ed.). (2004). *Oxford studies in metaphysics* (Vol. 1). Oxford University Press.

Zimmerman, D. (2005). The A-theory of time, the B-theory of time, and "taking tense seriously". *Dialectica, 59*, 403–459.

Zimmerman, D. (2008). The privileged present: Defending an "A-theory" of time. In J. Hawthorne, T. Sider, & D. Zimmerman (Eds.), *Contemporary debates in metaphysics* (pp. 211–225). Blackwell.

Zimmerman, D. (2011). Presentism and the space-time manifold. In C. Callender (Ed.), *The Oxford handbook of philosophy of time* (pp. 163–244). Oxford University Press.

The manufacturer's authorised representative in the EU is Springer Nature Customer Service Centre GmbH, Europaplatz 3, 69115 Heidelberg, Germany. If you have any concerns regarding our products, please contact ProductSafety@springernature.com

Printed and bound by CPI Group (UK) Ltd, Croydon, CR0 4YY
30/09/2026
02225460-0015